WASTE AND WASTE MANAGEMENT

WASTEWATER TREATMENT

PROCESSES, USES AND IMPORTANCE

WASTE AND WASTE MANAGEMENT

Additional books and e-books in this series can be found on Nova's website under the Series tab.

WASTE AND WASTE MANAGEMENT

WASTEWATER TREATMENT

PROCESSES, USES AND IMPORTANCE

ADRIANA MAGDALENA
EDITOR

nova science publishers
New York

Library of Congress Cataloging-in-Publication Data

ISBN: 978-1-53616-370-4

Published by Nova Science Publishers, Inc. † New York

CONTENTS

Preface		**vii**
Chapter 1	Wastewater Treatment and Agricultural Uses *Sufia Irfan and Basma M. Alharbi*	**1**
Chapter 2	Performance of Municipal Wastewater Treatment Procedures *Ibrahim Chourouk, Hammami Salah and Hassen Abdennaceur*	**31**
Chapter 3	Exploring the Energy Potential of Wastewater with Microbial Fuel Cells *Brandon E. Oliphant, Stephen A. Caponetti, Pauline Sow, Jessica Boyer, Shivani Amin, Lauren Bahnsen and Birthe V. Kjellerup*	**69**
Chapter 4	Treatment and Uses of Biogas and Bioethanol Wastewater *Fábio Spitza Stefanski, Thamarys Scapini, Aline Frumi Camargo, Caroline Dalastra, Natalia Klanovicz, Karina Paula Preczeski, Fabiane Czapela, Simone Kubeneck, Gislaine Fongaro and Helen Treichel*	**95**

Chapter 5 Petroleum Wastewater Treatment Using Granular Sequencing Batch Reactor: Physical Characteristics and Capabilities of the Aerobic Granules **145**
Shabnam Taghipour and Bita Ayati

Chapter 6 Improving Biomethane Production by the Anaerobic Co-Digestion of Agro-Industrial Wastes (Vinasse, Whey and Glycerin) Using Sequencing Batch Biofilm Reactors **181**
R. Albanez, G. Lovato, J. N. Albuquerque, S. P. Sousa, S. M. Ratusznei and J. A. D. Rodrigues

Chapter 7 Reclamation and Reuse of Grey Water: A Beneficiary Solution for the Water Demand Issues **201**
Rajagopalan Varadarajan, Manjula Gopinathan and Abirami Mani

Index **221**

Related Nova Publications **225**

PREFACE

Wastewater Treatment: Processes, Uses and Importance begins by providing information about wastewater treatment and its various application, especially in agriculture sectors. Some information about wastewater use and regulation in Saudi Arabia is also discussed.

The main characteristics of the natural oxidizing pond system and the activated sludge procedure are described and their performance in the abatement of physico-chemical, bacteriological and virological pollution is discussed.

Next, the authors describe and discuss the most common wastewater treatment processes and the importance of the activated sludge process in wastewater treatment, as well as introduce the idea of implementing microbial fuel cells into the procedural design of wastewater treatment for resource recovery.

This compilation also covers strategies aimed at minimizing the expense of water during every stage of energy and biofuel production, as well as forms of reuse and recycling that guarantee the utilization of wastewater in order to develop the circular economy in biogas and bioethanol plants.

Characteristics of cultivated bio-granules including the number of granules, density, sludge volume index, settling velocity, and physical strength are studied and their performance in treating petroleum and other types of wastewater is investigated.

Following this, the authors provide an overview of the achievements of studies in which anaerobic sequencing batch biofilm reactors have been used to co-digest agro-industrial wastes for the production of methane, with a focus on operational strategy and perspectives for energy estimations.

The closing study discusses the characteristics of grey water and available methods for its recycling and reuse. Grey water is the wastewater from homes, excluding black water, which typically makes up 50 to 80% of wastewater.

Chapter 1 - In residential, commercial and industrial communities, organizations are involved in different activities that make the water extremely polluted, and therefore, it has no use for other functional purposes. Since wastewater comes from different sources, it has a combination of suspended and dissolved organic and inorganic matters, soaps, detergent, fertilizers, plastic products, inorganic and organic compounds, and chemicals. Treatment of wastewater is done to save the ecosystem and people from the adverse outcomes of water pollution such as lack of dissolved oxygen, which can cause deterioration of the aquatic environment for plants and animals, excessive nitrogen and phosphorus in the water can cause ecosystem eutrophication, contamination of useable water, restrictions on recreational water activities etc. The rapid expansion of urban populations, the increasing domestic water need and the construction of city drainage systems are producing a considerable quantity of municipal waste and sewage water. Awareness of environmental safety and water pollution problems is one of the great needs of the time, and already, global environmental organizations are focusing on the safe and advantageous disposal of wastewater. Among various applications for wastewater, the agriculture sector has proven to be a functional area where treated wastewater can be an alternative to agriculture irrigation, especially in arid and semi-arid geographical regions. However, compared to the substantial agricultural industry of other countries, treated wastewater applications as a source of irrigation can be a small proportion, despite that it can aid in the conservation of water and its consumption in other sectors as well. Treatment of wastewater eliminates the suspended solids from the water so that the remaining effluent can flow out to the environment. Wastewater treatment management and supply operations into an agricultural water resource and land planning can be an economical and effective measure to minimize the severity of water pollution and controlling water pollution and consuming the sewage nutrients as crop essential growth compounds. The sewage treatment process has many significant steps to ensure desired sewage disposal for the safety of the environmental and human health. The first stage of wastewater treatment involves primary sedimentation where solid particles of organic matter are expelled by a gravity settling procedure. The primary treatment process, removes approximately 60% of suspended solids from the treated water, as the decay of solid substances consumes oxygen, which is also a necessity of aquatic plants and animals. The activated sludge process of wastewater involves the second stage where natural microorganisms break down the dissolved and suspended organic compounds. In the primary treatment, the aeration method returns oxygen into the wastewater. The secondary treatment of wastewater expels about 90% of the suspended solids. Tertiary treatment considers the use of chlorine to get rid of any waterborne pathogens and bacteria that could be a risk factor for human health. This chapter provides information about wastewater treatment and its various applications, especially in agriculture sectors. Some information about wastewater use and regulations in Saudi Arabia are also discussed.

Chapter 2 - Wastewater biological treatment is one of the most common kinds of treatments used to improve the physico-chemical and microbiological quality of wastewater in the world for recycling, agricultural reuse and safe release in the natural receiving environments; mainly in developing countries and in small towns. The natural oxidizing pond system and the activated sludge procedure are the oldest and the most widely biological wastewater treatment ones in the world intended for the sewage treatment. However, the implementation of the rotating biological disks and the trickling filter procedures were recently carried out in some wastewater treatment plants (WWTPs). The objective of this chapter is to describe the main characteristics of these procedures and to evaluate their performance in the abatement of physico-chemical, bacteriological and virological pollution. In addition, the various wastewater treatment steps, the microbial communities and the mechanisms responsible for the microorganism elimination in these procedures were reviewed.

Chapter 3 - Wastewater treatment is an essential process that ensures the public health by promoting the health and safety of water systems. In the most general sense, wastewater treatment is the combination of processes like biological transformation, filtration, sedimentation, and disinfection, which are used to purify the wastewater. This chapter describes and discusses the most common wastewater treatment processes, the importance of the activated sludge process in wastewater treatment, and finally introduces the idea of implementing microbial fuel cells (MFCs) into the procedural design of wastewater treatment for resource recovery. Microbial fuel cell technology is a wastewater treatment approach that uses microorganisms to convert chemical energy stored in the organic components of the wastewater into electrical energy. MFCs treat wastewater by taking advantage of respiring bacteria's ability to digest organics and produce free electrons. Current wastewater treatment approaches require significant energy input, however during these processes, the energy that is being transformed within the system is not captured but is instead released as waste energy (i.e., heat). This chapter reviews the feasibility of implementing microbial fuel cell technology into current wastewater treatment approaches and introduces the steps that need to be taken to utilize the electric energy released by bacteria and thus appropriately design the MFC process. Depending on the type of wastewater being used, an MFC can be customized by choosing from a variety of anodes, cathodes, and cation specific membranes which can, in turn, promote energy production. MFCs have the potential to not only provide a sustainable wastewater treatment approach but they can also serve as alternative processes that are energetically beneficial to the user and specific to a variety of wastewater types, geographical locations, and environments. Therefore, this chapter concludes with a discussion on possible designs of industrial MFCs and comments on their respective benefits and drawbacks, in regard to industrial implementation.

Chapter 4 - Renewable fuels, such as biogas and bioethanol, represent main strategies in the energy sector to replace dependence on fossil fuels. However, both processes link the production of biofuels, as well as combine with the formation of wastewater inherent to production stages that affect the concepts of green economy and pose risks to the environment. In this way, post-treatment processes prior to release in nature are required to ensure the depollution of the wastewater. In this sense, this chapter seeks to cover strategies aimed at minimizing the expense of water during every stage of energy and biofuel production, as well as forms of reuse and recycling associated with treatments that guarantee the utilization of wastewater in order to develop the circular economy in biogas and bioethanol plants finding ways for a more eco-friendly energy production.

Chapter 5 - In general, petroleum wastewater has high concentration of organic and inorganic components such as BTEX, MTBE, PAH, phenol, ammonia, sulfides, cyanides, and heavy metals which make it necessary to be treated before releasing to the environment. Granular sequencing batch reactor is one of the promising biological methods for cultivating aerobic granules and consequently treating wastewater. Owing to numerous advantages (i.e., stronger microbial structure of the granules, high removal efficiency, sludge dewatering ability, appropriate settle-ability, and high biomass maintenance and resistance against organic loading shocks), this technology has been used in remediation of various pollutants from aqueous solutions. In this chapter, characteristics of the cultivated bio-granules including number of granules, density, sludge volume index, settling velocity, and physical strength have been studied and their performance in treating petroleum and other types of wastewater have been investigated.

Chapter 6 - Anaerobic digestion is a promising technology for wastewater treatment as it allows the recovery of energy from biogas (hydrogen and/or methane) generated as a final product of the process. In addition to its low-cost operation, it significantly reduces the polluting load of the liquid residue. However, biogas production via anaerobic digestion depends greatly on substrate composition, such as fraction of organic matter, as well as on the amount of nutrients available for microorganisms. In this context, anaerobic co-digestion, consisting of the simultaneous anaerobic digestion of multiple organic residues in a single digester, is an interesting technique to improve organic waste digestion and its conversion to biogas. The agro-industrial wastes vinasse (sugarcane stillage), whey and glycerin usually present medium or low process yields when digested alone, making their co-digestion with other residues, produced in large quantities and with complementary characteristics, an attractive option. Therefore, a mini review has been performed regarding studies that dealt with the co-digestion of these wastes (vinasse with whey, vinasse with glycerin and whey with glycerin) for methane production. In these co-digestion processes an anaerobic sequencing batch (or fed-batch) biofilm reactor (AnSBBR) was used, which is one of the several high-rate configurations used as an alternative to continuous systems. Advantages of the AnSBBR include better effluent control and simple operation consisting of three stages: feed, reaction and discharge. AnSBBRs have been applied to the treatment of vinasse (bioethanol production), whey (dairy industry) and glycerin (biodiesel production) using various operational strategies: feeding mode, temperature, organic load, influent concentration and cycle length. Thus, this review presents an overview of the achievements of studies in which AnSBBRs have been used to co-digest agro-industrial wastes for the production of methane, with the focus on operational strategy and perspectives for energy estimations.

Chapter 7 - As per the world water development report the global demand for water has been increasing at rate of 1% for every year due to population growth, economic development and changing consumption pattern. Increasing the demand of water due to exponential growth of population led to the idea to recycle of waste water. In Wastewater treatment greywater recycling is rising as an integral part of water demand management, contributing to the progress of the preservation of high quality fresh water and also reducing pollutants in the environment and lowering the overall supply costs. An alternative source for treasured potable water is grey water. Grey water is the waste water from homes excluding black water. Typically 50 to 80% of waste water is grey water. This chapter will discuss about the character of grey water and available methods to recycle and reuse grey water.

In: Wastewater Treatment
Editor: Adriana Magdalena
ISBN: 978-1-53616-370-4

Chapter 1

WASTEWATER TREATMENT AND AGRICULTURAL USES

Sufia Irfan** and Basma M. Alharbi
Biology Department, College of Science
Faculty of Science, University of Tabuk
Tabuk, Kingdom of Saudi Arabia

ABSTRACT

In residential, commercial and industrial communities, organizations are involved in different activities that make the water extremely polluted, and therefore, it has no use for other functional purposes. Since wastewater comes from different sources, it has a combination of suspended and dissolved organic and inorganic matters, soaps, detergent, fertilizers, plastic products, inorganic and organic compounds, and chemicals. Treatment of wastewater is done to save the ecosystem and people from the adverse outcomes of water pollution such as lack of dissolved oxygen, which can cause deterioration of the aquatic environment for plants and animals, excessive nitrogen and phosphorus in the water can cause ecosystem eutrophication, contamination of useable water, restrictions on recreational water activities etc.

---

* Corresponding Author's E-mail: sirfan@ut.edu.sa.

The rapid expansion of urban populations, the increasing domestic water need and the construction of city drainage systems are producing a considerable quantity of municipal waste and sewage water. Awareness of environmental safety and water pollution problems is one of the great needs of the time, and already, global environmental organizations are focusing on the safe and advantageous disposal of wastewater. Among various applications for wastewater, the agriculture sector has proven to be a functional area where treated wastewater can be an alternative to agriculture irrigation, especially in arid and semi-arid geographical regions. However, compared to the substantial agricultural industry of other countries, treated wastewater applications as a source of irrigation can be a small proportion, despite that it can aid in the conservation of water and its consumption in other sectors as well. Treatment of wastewater eliminates the suspended solids from the water so that the remaining effluent can flow out to the environment. Wastewater treatment management and supply operations into an agricultural water resource and land planning can be an economical and effective measure to minimize the severity of water pollution and controlling water pollution and consuming the sewage nutrients as crop essential growth compounds.

The sewage treatment process has many significant steps to ensure desired sewage disposal for the safety of the environmental and human health. The first stage of wastewater treatment involves primary sedimentation where solid particles of organic matter are expelled by a gravity settling procedure. The primary treatment process, removes approximately 60% of suspended solids from the treated water, as the decay of solid substances consumes oxygen, which is also a necessity of aquatic plants and animals. The activated sludge process of wastewater involves the second stage where natural microorganisms break down the dissolved and suspended organic compounds. In the primary treatment, the aeration method returns oxygen into the wastewater. The secondary treatment of wastewater expels about 90% of the suspended solids. Tertiary treatment considers the use of chlorine to get rid of any waterborne pathogens and bacteria that could be a risk factor for human health. This chapter provides information about wastewater treatment and its various applications, especially in agriculture sectors. Some information about wastewater use and regulations in Saudi Arabia are also discussed.

Keywords: agriculture, nitrogen, phosphorus, Saudi Arabia, sludge, wastewater

INTRODUCTION

Arid and semi-arid zones of the world are facing a severe problem of water scarcity, so their focus is on any alternate source of water, which can have an economical and practical purpose in agriculture and the development sector. Rapid population expansion has caused an increased demand for food production. Irrigation has been long recognized for agriculture purposes and to develop rural society. Using wastewater in agriculture could be an essential measure in arid and semi-arid regions. However, wastewater treatment planning strategy should consider that the amount of wastewater in many countries would only account for a small proportion of total irrigation water coverage.

Nevertheless, using treated wastewater gives a satisfactory outcome in the conservation of high-standard water and its use for several other purposes. Rural regions with little freshwater resources can incorporate wastewater into agricultural irrigation and land designing, and this could be a cost-effective alternative to help with the financial burden. The strategic plan of municipal wastewater usage helps to control surface water pollution problems, conserves freshwater resources and sewage nutrients (Irfan and Shardendu, 2009) and organic wastes present in the (Senesi, 1989) can help grow healthy crops. The nitrogen and phosphorus content of sewage water might work as a replacement for commercial fertilizers (Figure 1). Sewage waste is an alternative source of bioenergy production, organic fertilizer, and also decreases the demand for freshwater consumption in the agriculture sector.

During the years 1550 and 1700, Germany, Scotland, and England were the first countries to utilize wastewater in agriculture irrigation (Drechsel et al. 2010; Tzanakakis et al. 2014). About 17% of the world's land has an irrigation water supply, which covers about 34% of the world's agriculture produce. The Middle East countries of the world such as, Arabian Peninsula, Egypt, Lebanon, Cyperus, Jordan where 30% of the cultivated area is under irrigation control, produces approximately 75% of the crops. These nations have a dependence on imported food products from other countries as well because of exceeding food demand. Australia and the USA have functional water resource planning for their extreme arid areas where wastewater is used for agricultural purposes while the high-quality water is used for drinking and other everyday applications. There have been plenty of wastewater reuse projects in the United States of America (USA). In 1912, Bakersfield, California used raw sewage as a source of irrigation for the first time (Pettygrove and Asano, 1985). According to the report, reclaimed water supply (16.9 million gal/day) from Bakersfield was used to irrigate almost 2,065 ha (5,100 acres) of corn, alfalfa, cotton, barley, and sugar beets produce (Pettygrove and Asano, 1985). In Florida, the city of Tallahassee has been using treated wastewater for agricultural farmland irrigation since 1966; this provision has been utilizing (18 million gal/day) of secondary effluent to water about 700 ha of land (Roberts and Bidak, 1994).

China has started using sewage water in agriculture since 1958, and now more than 1.33 million hectares of land are covered by sewage effluent irrigation. Middle Eastern countries like Jordan and the Kingdom of Saudi Arabia follow a national policy to reutilize treated wastewater effluents and are already completing this goal.

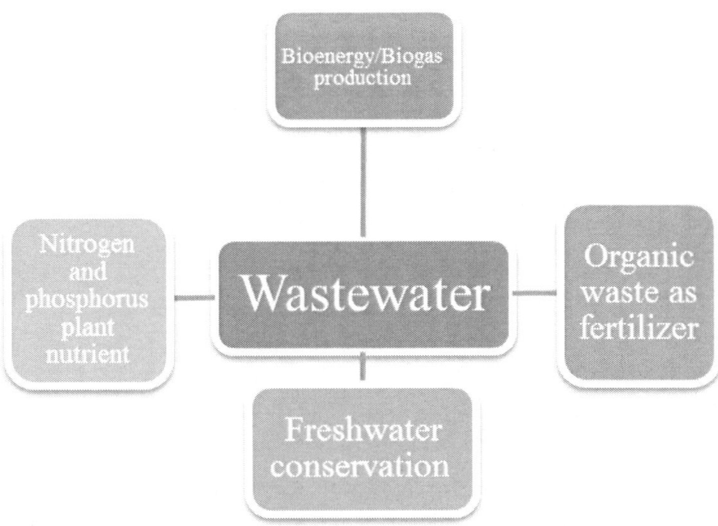

Figure 1. Illustration of various uses of wastewater.

It is acceptable to use treated wastewater for agriculture, considering the agronomic and economic aspects, but proper attention and safety measures are needed to minimize the adverse effects on the health of humans and the environment. A WHO (1989) Technical Report on 'Health Guidelines for Practical Use of Wastewater in Agriculture', suggested a combination of various safety measures to ensure the population's health protection. Some countries have strict legal systems for environmental protection, hampering the practical benefits of the method, whereas in other countries skilled technical staffs are not proficient enough to perform the wastewater treatment procedure. 'Agriculture' water is categorized according to its usage to define the advantages and disadvantages in order to gain the maximum benefit in crop production. The application of water in the field for irrigation purposes involves various steps mentioned in the user's guidelines, which also requires adjustments in the water application per field conditions such as appropriate climate conditions, physical and chemical properties of the soil, soil salinity of the selected crop and management procedures. Hence, treated wastewater has common characteristics suitable for use in an average field condition (WHO, 2006). According to the worldwide research report, vegetable and cereal crops can be irrigated using diluted or raw wastewater (Jaramillo, 2017; Qadir et al. 2010). Since vegetable crops grow in direct contact with water and are eaten in raw form so maintenance of crop restrictions can help in decreasing human health hazards.

In the city of Aleppo in Syria only 7% of agricultural lands growing vegetable crops are under wastewater irrigation coverage because of strict government vigilance (Qadir et al. 2010). However, high demand for vegetables in cities sometimes keep the restrictions at bay and it is difficult to maintain the agricultureal control, but only some agricultural products help farmers earn profit from their cultivation (Drechsel et al. 2002). The best alternative is to use wastewater in the irrigation of the agroforestry sector for the trees grown for fuel and timber uses (Minhas and Samra, 2004).

Table 1. Major constituents of typical domestic wastewater

Constituent	Concentration (mg/l)		
	Strong	Medium	Weak
Total solids	1200	700	350
Dissolved solids (TDS)	800	500	250
Suspended soild	350	200	100
Nitrogen	85	40	20
Phosphorus	20	10	6
Chloride	100	50	30
Alkalinity	200	100	50
BOD_5	300	200	100
Grease/oil etc.	150	100	50

Source: UN Department of Technical Cooperation for Development (1985), Mara, (1976).

In sewage and municipal wastewater, 99.9% contains a relatively small amount of suspended and dissolved organic and inorganic solids. Carbohydrates, lignin, fats, soaps, synthetic detergents, proteins, and their decaying components are organic wastes. Other natural and synthetic organic chemicals present in the sewage effluents are present in industrial wastewater. In Florida, and other semi-arid regions of the USA, the practice of using reclaimed wastewater in crop irrigation has been in effect for a long time. Wastewater use as a source of irrigation depends on the chemical factors and availability of nutrients such as nitrogen and phosphorus-fertilizers promoting healthy crop growth (Westcot and Ayers, 1985). Significant components of strong, medium and weak domestic wastewaters are presented in Table 1.

Figure 2 represents the procedure for sewage control and its stepwise treatment process. In primary treatment, the quiescent tank is used to store the sewage without any interference, to give time for heavy solids to settle to the bottom whereas other materials such as oil, grease, and light compounds float on the surface. The settled and floating materials and the remaining liquid flow out or are used for secondary treatment. Biological treatment of sewage involves the bacteria and other small organisms like nematodes to breakdown the organic waste via cellular respiration. Wastewater is rich in organic matter such as domestic garbage, kitchen waste, and broken food particles. Anaerobic digestion of sewage sludge breaks down the biodegradable materials with the help of bacteria in the absence of oxygen. Anaerobic digestion is an important step that produces a green source of energy in the form of marsh gas. Anaerobic digestion also supports renewable energy and works as an integrated waste management system, which helps the reduction of landfill gases into the atmosphere. The process also provides fertilizers for the energy crop's cultivation; maize, wheat and rice. At the final stage, liquid and solids separate from the sludge water, and water present in the sludge is reduced by the process of centrifugation, filtration, and evaporation. Sludge thoroughly passing through these mechanisms is not only economical for transportation but also suitable for composting.

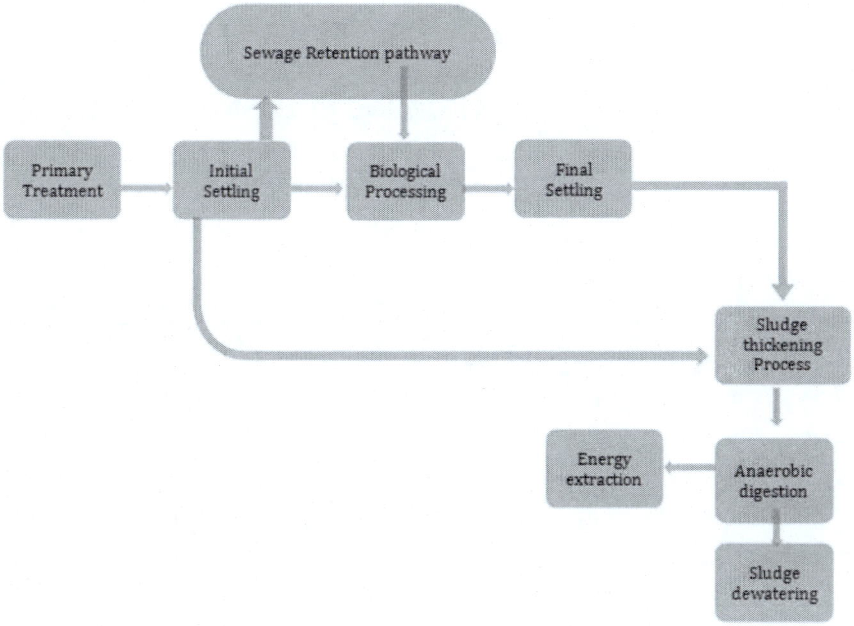

Figure 2. Sewage Retention Pathway

At this stage of dewatering, sludge becomes solid with 50-75% water content. Oxidation ponds or activated sludge methods are technologies used for the treatment of domestic wastewater, and such treatment methods need a substantial expenditure, so their practical application is difficult to implement. Sludge consistency varies from the slurry to dry solids. It has a mix composition of organic materials and nutrient enriched organic matters. The original use of sludge in agriculture as fertilizer was in a city in Ohio. The application of sludge as fertilizer was an original idea, and many researchers have studied its plant nutrient bulk (Rudolfs and Gehm, 1942; Sommers, 1977; Kirchmann et al., 2017). Baltimore, Maryland, has also applied excrement and other waste from the septic tank to crops (Allen, 1912). The nutrient content found in sludge is similar to organic waste such as animal excreta used for soil amelioration.

In addition to vital plant nutrients, sludge also contains trace elements that are essential for plant growth (Sensesi, 1989). Municipal sludge has loads of plant nutrients along with trace metals. Hence, the application of sludge on agriculture land helps to fix metal deficiencies (Logan and Chaney, 1983). Earlier, sludge was used as fertilizer without considering its detrimental effects on soil or crops (Allen, 1912). Since the 1970s, more significance was given to sludge application on farm land (Hinesly et al. 1972; Kirkham, 1974). Sludge's land application is widely acceptable in France, Spain, and the UK, but the practice is not approved in the Netherlands and Switzerland, where sludge is incinerated. In other European nations such as Greece, Malta, and Romania, using sludge as landfill is the prevalent solution. The Edmonton Compositing Treatment Station is the biggest sewage and sludge processing facility in the North America (Edmonton Composting facility, 2015). Many sewage treatment plants in NewYork City dewater the sludge with the help of extensive centrifugation along with a polymer to further dewater the sludge. The final output is called compost cake, which is collected by companies to change this compound into fertilizer pills. This biosolid is sold to regional farmers and is also used in turf farming as a fertilizer or soil amendment, which also reduces the requirement for sludge disposal area.

Backland people from the metropolitan region of southern California, return the sewage sludge to the sewer system of communities to process the compost at treatment plants situated on the Pacific coast. This process helps the local recycling of treated wastewater that can be used as an energy resource and reduces the required size of interceptor sewers and allows local recycling of treated wastewater while retaining the economy of a single sludge processing facility. This is an example of how sewage sludge can help solve an energy crisis (Altenergymag, 2017).

Table 2. The permitted limit for greywater reuse according to practical use

Test	Irrigation of ornamental fruit trees and fruit cultivation (A)	Irrigation of vegetables eaten uncooked (B)	Toilet use (C)
BOD5 (mg/l)	240	20	10
Sample number	Sample/month	Two samples/month	Sample/week
TSS	140	20	10
Sample number	Sample/month	Two samples/month	Sample/week
Thermotolerant coliforms (cfu/100 mL)	1000	200	10
Sample number	Two samples/month	Sample/two weeks	Sample/week

Source: Report on the WHO/AFESD regional consultation to review national priorities and action plans for wastewater reuse and management (WHO EM/CEH/106/E).

Municipal water discharge is posing serious threats to the animals and plants present in the ecosystem; therefore, the sewage needs proper treatment before release into the environment (Mulling et al. 2014). Activated sludge and oxidation pond technologies are used for domestic wastewater treatment but these practices are expensive (Shen et al. 2013). In recent years, a substitute system of engineered wetland ecosystems was accepted because of environment-friendly benefits. These wetland systems have rooted, emergent, and free-floating aquatic plants suitable for the wetland climate zone where bacterial interaction, nutrient uptake, and accumulation cleans the wastewater (Irfan and Shardendu, 2009). Another concept of wastewater treatment uses facultative ponds, where wastewater goes through the process of algae and bacterial cleaning treatment.

Table 3. Recommended maximum limit for trace elements in wastewater and their observed effects on crops

Elements	Recommended maximum limit (mg/l)	Comments
Al	5.0	Produces non-productivity in acid soils (pH < 5.5), but more alkaline soils at pH > 7.0 will precipitate the ion and eliminate any toxicity.
As	0.10	Wide toxicity variation in plants ranging from 12 mg/l for Sudan grass to less than 0.05 mg/l for rice.
Cd	0.10	Toxic to beans, beets and turnips crops at concentrations as low as 0.1 mg/l in nutrient solutions. Conservative limits recommended due to its potential for accumulation in plants and soils to concentrations that may be harmful to humans.
Co	0.05	Toxic to tomato plants at 0.1 mg/l in nutrient solution. Tends to be inactivated by neutral and alkaline soils.
Cr	0.10	Not recommended as an essential growth element. Conservative limits recommended due to lack of knowledge on its toxicity to plants.
Cu	0.20	Toxic to a number of plants at range 0.1 to 1.0 mg/l in nutrient solutions.
Fe	5.0	Not toxic to plants in aerated soils, but can contribute to soil acidification and loss of availability of essential phosphorus and molybdenum. Overhead sprinkling may result in unsightly deposits on plants, equipment and buildings.
Mn	0.20	Toxic to a number of crops at a few-tenths to a few mg/l, but only in acid soils.
Mo	0.01	Not toxic to plants at normal concentrations in soil and water. Can be toxic to livestock if forage is grown in soils with high concentrations of available molybdenum.
Ni	0.20	Toxic to a number of plants at 0.5 mg/l to 1.0 mg/l; reduced toxicity at neutral or alkaline pH.
Pd	5.0	Can inhibit plant cell growth at very high concentrations.
Sn		Excluded by plants, specific tolerance unknown.
Se	0.02	Toxicity to plants at concentrations as low as 0.025 mg/l and toxic to livestock if forage is grown in soils with relatively high levels of added selenium. As essential element to animals but in very low concentrations.
F	1.0	Inactivated in presence of neutral and alkaline soils.

Sources: Food and Agriculture Organization. Water quality for agriculture. Irrigation and Drainage Paper 29 Rev. [1]1985.

Such constructed wetland systems are controlled by vegetation, which plays an essential role in the treatment processes. In the free-floating plant communities, water hyacinth (*Eichhornia crassipes*) emerged as the best pollutant removal plant due to its speedy growth and extensive root system (Olukanni and Kokumo, 2013; Rezania et al. 2013). The water hyacinth lagoons are based on horizontal trickling function, where roots of submerged plants form physical transparent support for the bacteria to grow (Dhote and Dixit, 2009). However, the construction of aquatic systems, especially the water hyacinth treatment basins, did not receive much acceptance because the whole project needs a large land area and high capital investment.

In arid climates, sewage effluents have specific characters where extreme temperature and rapid evaporation in the presence of low humidity helps salt deposition in the soil profile. Certain features of soil such as particle dispersion, aggregates stability, soil structure, and permeability are easily affected by the type and concentration of exchangeable ions present in irrigation water. Free ions do not affect plant yield and are sometimes beneficial at lower concentrations, but may produce phytotoxicity at high concentrations in two ways (i) direct interruption in plant metabolism (ii) indirect effect on the growth nutrients and preventing their assimilation into plants. Morishita (1985) has reported that soil with Cd around 0.4-0.5 ppm has a rice yield with 0.08 mg Cd, while an increase in metal concentration up to 0.82, 1.25, 2.1 ppm of Cd has grown the rice with Cd contamination around 1 mg.

Municipal effluents may have toxicity in them because of heavy metals, which is coming from many small and unofficial industries discharging wastes into the conventional sewer drainage stations. These toxic elements are present in small amounts, hence, termed as trace elements. During the treatment method, some trace elements are removed, but others remain in the water and could cause phytotoxic effects. Thus, municipal wastewater effluents should be examined to find out the trace element toxicity and associated health hazards. The National Academy of Sciences (1972) and Pratt (1972) had proposed recommended concentrations of trace elements and their effects on crops' productivity (Table 3).

Table 4. Biological and chemical contaminations present in the raw wastewater used in agriculture

Biological	Bacteria*	*E. coli, Vibrio cholerae, Salmonella spp., Shigella spp.*
	Helminths*	*Ascaris, Ancylostoma, Tenia spp*
	Protozoans*	*Intestinal Giardia, Crysptospridium, Entamoeba spp.*
	Virus* Schistosoma**	Hepatitis A and E, Adenovirus, Rotavirus, Norovirus Blood-flukes
Chemical	Substanc of sanitary interest Heavy metals** Hydrocarbons** Pesticides*	As, Cd, Hg Dioxins, Furans, PCBs Aldrin, DDT

Contact* and/or consumption; **Consumption; Source: WHO (2006).

Table 4 listed the biological and chemical sources of health risks found in wastewater used for agricultural irrigation. The concentration levels and types of pathogens and chemical substances present in wastewater causes health risks when used in agriculture irrigation. These risky conditions depend on the sanitary and socioeconomic development of the communities (Gerba and Rose, 2013). A study published by Jiménez et al. (2010) has estimated the concentration of viruses, protozoan parasites and helminths in wastewater is 10–1000 times higher in developing countries compared to developed countries. Municipal wastewaters have organic concentrations in very small quantities and ingestion over prolonged periods can cause detrimental effects on human health. The health-associated problem begins with the chemical compounds of wastewaters, when they contaminate the growing crops or groundwaters.

According to the report, over 3300 wastewater treatment amenities were provided under the scheme of AQUAREC international project between the years 2000-2006. The wastewater facilities have different water treatment quality standards and purposes among which the agricultural sector is the primary user of wastewater applications (Wintgens et al. 2006). The maximum number of wastewater reuse systems available in Japan and the United States was 1800 and 800 respectively followed by Australia and Europe with 450 and 230 respectively. In the Mediterranean and Middle Eastern zone, almost 100 wastewater treatment facilities were established. In South America, 50 treatment systems were identified, whereas in the African Sahara region, 20 treatment structures were reported (Wintgens et al. 2006). According to a report published by FAO, only 10% of the total global agricultural land is untreated or partially treated with wastewater irrigation, which comprises about 20 million hectares in almost 50 countries (EPA, 2012).

The application of wastewater in irrigation to conserve water and to use the nutrients and other compounds present in wastewater as plant growth promoter has some severe issues as well. Many research reports and studies have associated the high prevalence of contamination in the countries where wastewater is the source of irrigation. Some studies assessed the epidemiological consequences in the population, but most of them did not consider it essential criteria to study the level of exposure in the population. Hence, complete research is needed to get awareness of the wastewater contamination or consumption of wastewater-contaminated food; this also includes unhygienic toilet conditions or lack of toilets for farmers and residents, and contamination of drinking water. Heavy metals are components of wastewater, the assessment of which requires financial support and advance scientific technology (Oden, 2019), which is still not well established in many countries. Some studies have presented the report combining water quality and its microbial risk assessment on the local population in developing countries.

Various studies have revealed that consumers who are using wastewater-irrigated crops are in primary health risk groups. In Accra, Ghana an urban street vegetable market that is selling fresh produce to meet the demands of local people is contaminated with wastewater. The reason behind this is that urban populations are not aware of the health risks of consuming these vegetables and are putting their lives at risk (Obuobie et al., 2006; Amoah et al. 2007).

Pathogens are a permanent resident of wastewater, but the presence of chemical toxicity of the industrial effluents cannot be ignored, especially in those countries where the industrial revolution has started more recently. These industrial pollutants enter aquatic systems and are biologically magnified through the food web. Leafy crops tend to accumulate a high concentration of metals such as Cadmium and Chromium. The trend of metal concentration in plants varies from being highest in roots and lowest in shoots and leaves. The more metal concentration in industrial effluents, the higher the concentration in plant tissues. Treatment of industrial effluent removes heavy metals and other toxic materials, but if the high concentration of metal still exists in reclaimed wastewater, the irrigation practice using treated water must not be encouraged because of hazardous effects of metals on agriculture crops which will ultimately affect the local people eating those food crops.

Emerging contaminants (ECs) are designated as molecules exhibiting variation in biological activity on different organisms. Their physicochemical characters help their persistence in the environment and aid their bioaccumulation. ECs are present in irrigated wastewater; their examples include analgesics, antihypertensive and antibiotics, which may produce human health hazards. In addition some ECs leads to endocrine disruption (ED) (Jackson and Sutton, 2008). These complex molecules were recently categorized as environmental contaminants because earlier, no scientific information was reported about their accumulation in soil, water, air and biological tissues. European countries such as Spain, Italy, Germany, Greece and France, along with Brazil from South America have approximated the release of almost 500 tons of analgesics, especially salicylic acid and diclofenac into surface water sources (Heberer, 2002). ECs are entering into aquatic resources through various anthropogenic actions, which thereupon create toxic residues and have adverse impacts on aquatic life and, finally on humans (Kasprzyk-Hordern et al. 2009). Municipal effluents coming from the treatment sewage plants are one of the primary EC resources, as conventional treatment plants cannot prevent the exposure of these compounds into the environment (Jackson and Sutton, 2008). Other anthropogenic contributors of ECs are pesticides and antibiotics emerging as dispersed pollutants from agriculture and farming sectors. In addition, water resources are contaminated by runoff from soil mixed with animal waste or sludge applied in the form of fertilizers.

ECs may also contaminate groundwater due to landfill leachate, exposure of manure storage or from drip irrigation with treated or untreated wastewater on agricultural land (Grassi and Farina, 2013). The effects of ECs on human systems are not well explained. However, these compounds have a role in the disruption of endocrine and immunological systems of aquatic organisms (Fent et al., 2006).

WASTEWATER USAGE IN THE KINGDOM OF SAUDI ARABIA

In KSA, the total water consumption is estimated at around 25 billion m^3 per year. The agriculture sector has 88% of the water usage while industries consume around 3% (Khodran and Baig, 2011; AlZahrani and Elhag, 2003). The total public sector consumption is about 9%, which is about 2.1 billion m^3 per year (AlZahrani and Elhag, 2003).

Following is the graphic illustration of standards for treated wastewater use prepared in 2006, which shows the minimum treated wastewater quality requirements for restricted (Figure 3) and unrestricted use (Figure 4) respectively in irrigation water (MWE: Ministry of Water & Electricity, 2012).

Business and industrial organizations of the Royal Commission situated in Jubail and Yanbu have approved treated wastewater guidelines where some of the values of the parameters are more than the KSA required standards such as turbidity, fecal coliform, and total dissolved solids as described above in Table 5.

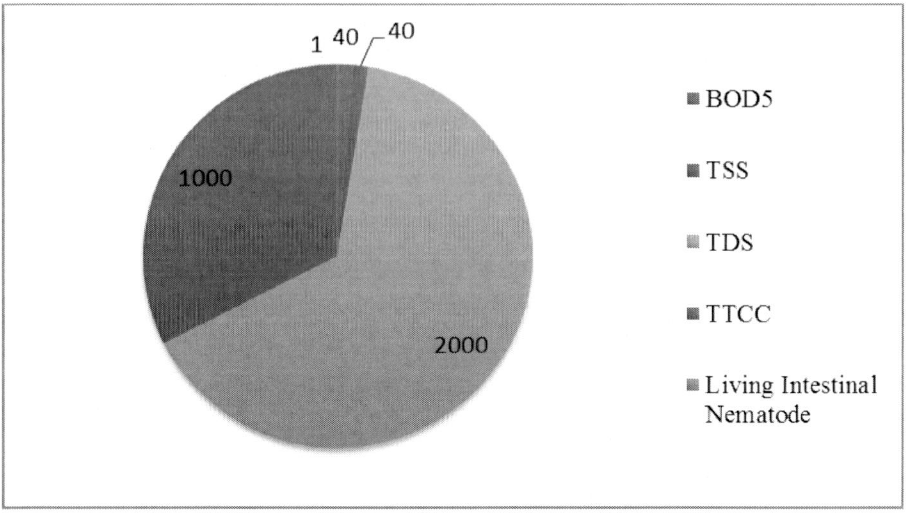

*TSS- Total suspended Solids *TDS- Total Dissolved Solids *TTCC- Total Thermotolerant Coliform Count.

Figure 3. Treated wastewater standards for restricted irrigation uses in Saudi Arabia

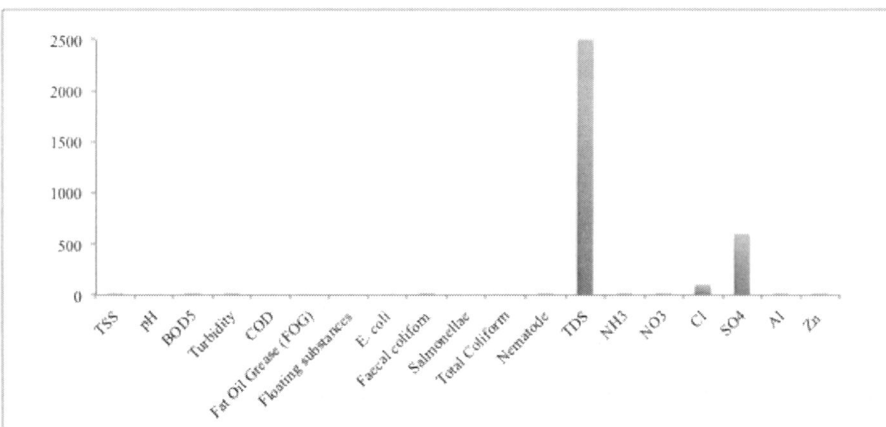

Units- mg/l

Figure 4. Treated wastewater standards for unrestricted irrigation use in Saudi Arabia

Table 5. Showing treated wastewater standards for use in unrestricted irrigation in the Kingdom of Saudi Arabia (KSA)

Characters	KSA standard	Royal Commission of Jubail and Yanbu standard
Turbidity (NTU)	5	2
COD	-	50
Faecal Coliform	2.2	1
Total coliform	nil	2.2
TDS	2500	1750
Cl	0.2-0.5	0.5
Boron	0.75	0.075
Chlorides	nil	500

Source: KAUST: King Abdullah University of Science & Technology (2011) and RCJY (2010).

REGULATIONS FOR TREATED WASTEWATER USE

In the year 1970, use of treated wastewater in KSA emerged as a wise idea, which was later encouraged by the Council of Leading Islamic Scholars of KSA, who issued a fatwa (Islamic declaration) for practical application of treated wastewater in KSA, which enabled the considerable use of treated wastewater in various sectors (MWE: Ministry of Water & Electricity, 2012). The Shoura Council, the group of Leading Islamic Scholars, instructed avoiding using wastewater for drinking purposes for the safety of human health. The first directive on treatment and reuse of wastewater was introduced in KSA in May 2000, under the supervision of 'Treated Sanitary Wastewater and Its Reuse Regulations'; with secondary or tertiary wastewater treatment levels guidelines (MWE: Ministry of Water & Electricity, 2012). After six years, in 2006, MWE has published two booklets 'Design Guidelines for Wastewater Treatment Plants in Saudi Arabia' and 'Using Treated Water for Irrigation: Controls, Conditions, Offenses and Penalties' (MWE: Ministry of Water & Electricity, 2012). These strategies have fulfilled the purpose of establishing wastewater treatments that follow safe methods and standard reuse practices.

The Kingdom of Saudi Arabia has a target to accomplish 100 percent utilization of treated wastewater by the year 2025. Achieving this development plan may place the Kingdom of Saudi Arabia as the third largest water reclamation and reuse nation in the world. Saudi Arabia has a target of reusing over 65% percent of water generated via anthropogenic wastes by 2020 and aiming for 90% wastewater reclamation by 2040. Saudi Ministry for Water and Electricity has an investment plan of $53 billion for the next 15 years for the sewage sector development. This investment plan will cover the wastewater treatment and water reclamation development under the statutory regulations for industry standards and the National Water Company (2008). The Saudi Water Act (2010) and the Treated Sewage Effluent (TSE) board have provided the Kingdom's agricultural and industrial division of treated wastewater a financially sustainable long-term market for treated sewage effluent usage. The Kingdom of Saudi Arabia has proposed a commission on National Water Strategy to tackle the water challenges in the Kingdom. Recently, 30-40 percent of reclaimed wastewater is in the reuse process, and this water reuse can further reach up to 241 million m^3 every year, and most of it is consumed in agriculture followed by landscape gardening, industrial and recreational applications. In the year 2010, about 240 million m^3 (almost one-quarter) of treated sewage water was used in landscape gardening for irrigation purposes in municipal and city parks in Riyadh, Al Taif, Jeddah, Dhahran, and Jubail (Ouda, 2014). Riyadh city was ranked highest in the utilization of almost 50% of treated wastewater, which is estimated to be about 120 million m^3.

The remarkable use of treated wastewater was seen in Hanifa Valley in the east of Riyadh city for recreational and ecological activities where treated wastewater was used to develop natural park areas for aesthetic green and healthy environment. This project also promoted the development of residential colonies, farming, and recreation pursuits along with cultural and tourism promotion among the urban population. Approximately 9000 hectares of agriculture lands were irrigated to support date palms and food crops cultivation. They used treated wastewater in various landscape activities of the urban ecosystem. Treated wastewater use in industrial and commercial centers of the city aims to expand further due to the high cost of desalinated water which costs almost USD 0.8/m^3

(Ouda, 2014; Ouda, 2013; Saline Water Conversion Corporation, 2010; 2012). In Al Kharj city, treated wastewater flows through a sewage waterway and is stored in a pond for filtration procedure using the sand particles filtration and finally to groundwater. The agriculture sector represents the biggest consumer of wastewater in KSA, which covered about 84% of water demand in the year 2010. In the KSA, The National Water Company has chosen treated wastewater as a leading business in terms of substantial financial profit, so it is targeting dealing with the consumers from industrial and other commercial divisions. Industrial and commercial zones were reported to claim 5% of the total wastewater (Ouda, 2015). The agriculture division also needs an integral scheme to utilize treated wastewater in irrigation to provide economic efficiency and sustainability.

By the end of 2035, there will be an estimated growth of about 2.5 km^3/y in treated wastewater collection. The main focus for treated wastewater utilization will be on its secure industrial application, which can bring high financial revenue. The targeted metro cities for the project implementation are Riyadh, Makkah, Madina, and Eastern Province regions (KAUST: King Abdullah University of Science & Technology, 2011; Ouda, 2015).

WASTEWATER FACILITIES IN KSA

The Ministry of Water and Electricity (MWE) has proposed a 100% wastewater system network in cities with populations of 5000 or more by 2025. Such a colossal project needs a generous expenditure in wastewater treatment (Drewes et al. 2012; KAUST: King Abdullah University of Science & Technology, 2011; MWE: Ministry of Water & Electricity, 2012). According to an estimation, around 6.8 million m^3/day (2.5 km^3/y) of treated wastewater will be available for use in the Kingdom by the end of government's Vision 2030 (Ouda, 2015).

Table 6. Presents the wastewater flow per region in the year 2010 and the projected wastewater flows for the years 2025 and 2035 with their flow capacity

Wastewater treatment plants										
	Calculated wastewater Flow (1000 m^3/day)			**Working**		**Under construction**		**Planned**		**Total planned Future capacity**
Province	2010	2025	2035	No	Cap	No	Cap	No	Cap	
Al Baha	56	81	107	0	0	1	16.2	6	72.6	88.8
Al Jouf	69	96	119	2	38	2	57.5	1	22	117.5
Asir	294	419	529	4	82.5	7	172.5	0	--	255
Eastern province	648	905	1128	13	527.3	6	251	1	400	1178
Hail	93	130	162	1	19.2	2	87.6	1	55.8	162.6
Jizan	203	301	381	1	20	0	--	22	12	132
Madinah	275	390	496	4	351	5	34	0	--	385
Makkah	1100	1532	1911	15	888	6	902	1	113	1903
Najran	78	107	133	0	0	1	60	5	170	230
Northern Borders	47	71	88	2	24	1	24	1	25	73
Qaseem	185	258	328	5	131	3	125	0	--	256.6
Riyadh	1,066	1,513	1,890	10	993	7	443.5	2	530	1967
Tabuk	120	169	214	1	60	1	15	0	--	75
	4234	5972	7486		3135		2188		1500.4	6823.8

Source: KAUST: King Abdullah University of Science & Technology (2011).

At present, wastewater treatment facilities are available in most of the cities in KSA (Table 6). The National Water Company has an ambitious project of wastewater treatment plants, which are under construction in many small and big cities. The highest numbers of treatment plant projects are in Riyadh (1967), followed by Makkah (1903) and Eastern province (1178). Northern Borders and Tabouk province of the Kingdom has the smallest number of the sewage treatment systems, which is 73 and 75, respectively (Saudi Geological Survey, 2012). Wastewater streams in Riyadh, Jeddah, Makkah, Al Taif, Madinah, and Dammam cities collect the water post-tertiary or secondary treatment. Conventional activated-sludge processing is the most prevalent secondary treatment technology in the KSA. For tertiary treatment, media filtration and chlorination disinfection are the accepted technologies. Urban cities also prefer to use advanced wastewater treatment mechanisms such as reverse osmosis. Sewage water after passing through a tertiary treatment process has various uses in reclamation purposes such as industries, agriculture irrigation, and landscape gardening.

Providing wastewater treatment facilities all over the cities is among agenda of the government's Vision 2030, which needs gross financial support. Therefore in December 2014, the Saudi Council of Ministers have proposed a new wastewater tariff for the government commercial and industrial divisions in December 2015. According to a report published in Arabnews (2014), the target sectors for the tax increment includes grand establishments where water consumption exceeds more than average household capacities such as hotels, private healthcare facilities, furnished residential apartments, malls and other commercial and industrial participants (Arabnews, 2014). The idea behind the introduction of the new levy was to promote treated wastewater utilization in the industrial and commercial zones for all nonpotable purposes and to support the long-term sustainability of water and wastewater infrastructure.

Wastewater Treatment Companies in Kingdom of Saudi Arabia

Almar Water Solutions, an associated group of Abdul Latif Jameel Company, has qualified for a wastewater treatment plant in West Dammam, Saudi Arabia. The project will begin with the treatment capacity of 300,000 cubic meters per day and will progress as a public-private participation (PPP) scheme. Almar Water Solutions has participated along with FCC Aqualia and Nesma companies. The project has a plan of a 25-year agreement for wastewater treatment, in which the National Water Company (NWC) will cover the sewage system.

The company, *Saudi Water Technology (SWT)* has an affiliation with Al Bassam group of companies and one of the developers in the water and wastewater treatment applications sector. They are also a leading supplier and wholesale distributor of a wide spectrum of water treatments systems, components and spare parts. SWT has completed many difficult water treatment projects owing to their up to date technological procedures.

ENTEC has solutions for water and wastewater treatment from design to operating. They have an array of custom technologies as per client needs. ENTEC focuses on meeting high-quality output with low cost using innovations in water and wastewater treatment.

Conclusion

The application of treated wastewater in the agriculture sector benefits society, the environment and the economy. This practice supports an alternative idea in different regions where urban society has developed with scarce water resources, especially groundwater resources influenced by climate variability and climate transformation. The most beneficial use of wastewater in agriculture irrigation is the control of water pollution. The cutback in municipal discharge will improve the source quality of collecting water (Toze, 2006). They will help in the preservation of groundwater basins, as they will get high quality water due to wastewater use in irrigation (Hernández, 2010). In addition to this, bulk use of wastewater could help in the well-established treatment plant to produce a quality stream to benefit irrigation water supply and sanitation projects (Hernández, 2010). In the suitable climatic and geographic regions, low-budget wastewater treatment systems might also be a viable option, where treated wastewater covers the demand of agricultural reuse. Wastewater use as an agricultural water supply helps to release the financial burden of the purchase of heavy expensive instruments. In some countries, wastewater reuse helps to reduce the municipal expenses of locating water sources in the region. Agricultural wastewater reuse provides the reasonable action in appropriate investments and financial policies for pollution mitigation and prevention.

Using raw or diluted sewage water is increasing in many developing countries of the world. Hence, government agencies have to face more challenges and responsibilities in the management of increasing availability and use of wastewater without causing a potentially adverse effect on public health and the environment. Wastewater collection and treatment technologies are limited so it is advisable for public organizations to consider preventive and safety measures that can reduce toxic contaminations from industrial effluents. We suggest vigorous supervision at an agricultural level to mitigate pathogen interaction between cultivated crops and farmers. Finally, post-harvest agricultural crop quality and public awareness for their protection from any apprehensive infection from agricultural produce also has a crucial role. Sustainable sanitation programs need considerable investment and a distinct approach to follow local standards and limitations that can benefit the society and ecosystem. Investment strategies should think carefully about the small-scale traders and low-income farmers whose livelihood depends on wastewater irrigation agriculture. KSA has achieved the highest rank in wastewater collection and treatment system plans, extensive coverage, and utilization of treated wastewater. The continued wastewater treatment proposal has achieved significant success, but many new schemes and proposals are compulsory to make it sustainable. The

Saudi government has taken effective action to achieve the guaranteed success in the development plan and its execution in the wastewater sector. Private sectors are also taken into consideration to reach great success in the implementation of the sufficient use of treated wastewater. The introduction of wastewater tariffs to the domestic sector can accompany the tariff process for the government, commercial, and industrialized divisions. This tariff scheme aims to administer monetary benefit to the wastewater system and encourage water conservation considering the social, cultural, and political scenario of the country. However, implementation of a wastewater tax to adjust the financial strain on the government's wastewater proposal might result in hostile public reaction because there is little awareness of the nation's water distribution and management plan. The levy on water and wastewater uses need regulations to maintain the services among the public and to secure feasibility of the system. Another approach calls for providing support to the competence of the public division, which can involve its private counterparts for the successful collaboration and continual development. Moreover, the public sector requires extensive human resources, government supervision, corporate participation, and treated wastewater database records.

References

Allen, K. (1912). Sludge Treatment in the United States. In *Sewage Sludge*. New York: McGraw Hill Book Co. 195-258.

Altenergymag. (2017) https://www.altenergymag.com/article/2017/03/can-sewage-solve-our-energy-crisis/25893.

Al-Zahrani, K. H. and Elhag, E.A. (2003) *Agricultural Development during the era of King Fahd*, Book published by King Saud University, Riyadh, KSA.

Amoah, P., Drechsel, P., Abaidoo, R. C. and Henseler, M. (2007). Irrigated urban vegetable production in Ghana: microbiological contamination in farms and markets and associated consumer risk groups. *Journal Water Health*, [5]:455–466.

Arabnews. (2014) *Residential users exempt from new tariffs on utilities.* Retrieved January 30, 2015, from http://www.arabnews.com/news/672311.

Dhote, S. and Dixit, S. (2009). Water quality improvement through macrophytes- A review. *Environment Moitoring Assessment*, [152]: 149–153.

Drechsel, P., Blumenthal, U. J. and Keraita, B. (2002). Balancing health and livelihoods: adjusting waste- water irrigation guidelines for resource-poor countries. *Urban Agriculture Magazine,* [8]: 7–9.

Drechsel, P., Scott, A., Sally, R., Redwood, M. and Bachir, A. (2010). *Wastewater Irrigation and Health: Assessing and Mitigating Risk in Low-Income Countries*; International Water Management Institute, Ed. Earthscan: London, UK.

Drewes, J. E., Patricio Roa Garduno, C. and Amy, G. L. (2012). Water reuse in the kingdom of Saudi Arabiastatus, prospects and research needs. *Water Science & Technology-Water Supply*, [12]: 926–936.

Edmonton Composting facility. 2015. *City of Edmonton*. Archived from the original on 26 March 2015.

Environmental Protection Agency. (2012). U.S. Agency for International Development. Guidelines for Water Reuse; U.S. Environmental Protection Agency: Washington, DC, USA.

FAO. (1985). Water quality for agriculture. R. S. Ayers and D. W. Westcot. *Irrigation and Drainage Paper 29 Rev*. 1. FAO, Rome. Page-174.

Fent, K., Weston, A. and Caminada, D. (2006). Ecotoxicology of human pharmaceuticals. *Aquatic Toxicol*ogy, [76]:122–159.

Gerba, C. and Rose, J. (2013). International guidelines for water recycling: Microbiological considerations. *Water Scence. Technol*ogy, [3]: 311–316.

Grassi, M., Rizzo, L. and Farina, A. (2013). Endocrine disruptors compounds, pharmaceuticals and personal care products in urban wastewater: Implications for agricultural reuse and their removal by adsorption process. *Environmental. Science Pollution Res*earch, [20]: 3616–3628.

Heberer, T. (2002). Tracking persistent pharmaceutical residues from municipal sewage to drinking water. *Journal Hydrology,* [266] :175–189.

Hernández, F., Molinos, M. and Sala, R. (2010). Economic valuation of environmental benefits from wastewater treatment processes: An empirical approach for Spain. *Science of Total Environ*ment, [408]: 953–957.

Hinesly, T. D., Jones, R. L. and Ziegler, E. L. (1972). Effects on corn by application of heatedanaerobically digested sludge. *Compost Science,* 13[4]:26-30.

Irfan, S. and Shardendu, S. (2009). Dynamics of nitrogen in subtropical wetland and its uptake and storage by *Pistia stratiotes*. *Journal Environemntal Biology*, 30 [6]: 977-81.

Jackson, J. and Sutton, R. (2008). Sources of endocrine-disrupting chemicals in urban wastewater, Oakland, CA. *Science of Total Environment,* [405]: 153–160.

Jaramillo, M. F., and Restrepo. I. (2017). *Wastewater Reuse in Agriculture:* A Review about Its Limitations and Benefits. Sustainability. 1-19.

Jiménez, B., Mara, D., Carr, R., and Brissaud, F. (2010). Wastewater treament for pathogen removal and nutrient conservation: Suitable systems for use in developing countries. In *Wastewater Irrigation and Health: Assessing and Mitigating Risk in Low-Income Countries*; Earthscan: London, UK.

Kasprzyk-Hordern, B., Dinsdale, R., and Guwy, A. (2009). The removal of pharmaceuticals, personal care products, endocrine disruptors and illicit drugs during wastewater treatment and its impact on the quality of receiving waters. *Water Res*earch, [43]: 363–380.

KAUST (King Abdullah University of Science & Technology). (2010, 2011). *KAUST industry and collaboration program (KICP), the KICP annual strategy study: Promoting wastewater reclamation and reuse in the kingdom of Saudi Arabia: Technology trend, innovation needs, and business opportunities*. Jeddah, KSA: KAUST.

Khodran, H. Al-Zahrani. and Baig, M. B. (2011). Water in the Kingdom of Saudi Arabia: Sustainable Management Option, *the Journal of Animal & Plant Sciences,* 21: pp. 601-603.

Kirchmann, H., Borjesson, G., Katterer, T. and Yariv, C. (2017). From agricultural use of sewage sludge to nutrient extraction: A soil science outlook. *Ambio* [46]: 143-154.

Kirkham, M. B. (1974). Disposal of Sludge on Land: Effects on Soils, Plants, and Ground Water. *Compost Science,* 15 [2]:6-10.

Logan, T. J., and Chaney, R. L.(1983). Metals. Pp. 235-323 in *Utilization of Municipal Wastewater and Sludge on Land*, A. L. Page, T. L. Gleason, J. E. Smith, I. K. Iskander, and L. E. Sommers, eds. Riverside: University of California.

Mara, D. D. (1976). *Sewage Treatment in Hot Climates*. John Wiley, London.

Minhas, P. S. and Samra, J. S. (2004). Wastewater Use in Peri-urban Agriculture: Impacts and Opportunities. Central Soil Salinity Research Institute, Karnal, India. Pages- 75.

Morishita, T. (1988). Environmental hazards o sewage and industrial effluents on irrigated farmlands in Japan. Ch. 6*, Treatment and Use of Sewage Effluent for Irrigation*. M. B. Pescod and A. Arar (eds). Butterworths, Severoaks, Kent.

Mulling, B. T. M., Soeter, A. M., van der Geest, H. G. and Admiraal, W. (2014). Changes in the planktonic microbial community during residence in a surface flow constructed wetland used for tertiary wastewater treatment. *Science of Total Environ*ment, [466-467]: 881–887.

MWE (Ministry of Water and Electricity). (2012). *Supporting documents for King Hassan II great water prize*. Riyadh, KSA. Retrieved November 30, 2014, from http:// www. world water council. org/ fileadmin/ wwc/ Prizes/ Hassan_ II/ Candidates_ 2011/ 16. Ministry_ SA.pdf.

National Academy of Sciences and National Academy of Engineering. (1972). Water quality criteria. US Environmental Protection Agency, Washington DC. Report No. EPA-R373-033. Page- 592.

National Water Company. (2008). *Management contract for water supply and wastewater collection services in the city of Jeddah, Kingdom of Saudi Arabia*. Riyadh, KSA: National Water Company.

Obuobie, E., Keraita, B., Danso, G., Amoah, P., Cofie, O. O., Raschid Sally, L. and Drechsel, P. (2006). *Irrigated urban vegetable production in Ghana: characteristics, benefits and risks.* IWMI-RUAF IDRCCPWF, Accra, Ghana: IWMI, Page-150. www.cityfarmer.org/ GhanaIrrigateVegis.html.

Oden, M. K. (2019). Evaluation of water quality in the Kizilirmak River (Central Anatolian) using physicochemical parameters and process proposal for improvement in water quality-a case study. *Desalination and Water Treatment,* [137]: 226-233.

Olukanni, D. O. and Kokumo, K. O. (2013). Efficiency assessment of a constructed wetland using *Eichhornia crassipes* for wastewater treatment. *American Jornal Engineering Res*earch,[2] :450–454.

Ouda, K. M. Omar. (2013). Towards Assessment of Saudi Arabia Public Awareness of Water Shortage Problem. Resources and Environment *International Journal of Water Resources Development,* 3 [1]: 10-13.

Ouda, K. M. Omar. (2015) Treated wastewater use in Saudi Arabia: challenges and initiatives. *International Journal of Water Resources Development* 32 [5]: 799-800.

Ouda, O. K. M. (2014). Domestic water demand in Saudi Arabia: Assessment of desalinated water as strategic supply source". *Desalination and Water Treatment*, 1–11. doi: 10.1080/19443994. 2014.964 332.

Pettygrove, G. S. and Asano, T. (1985). Eds. *Irrigation with Reclaimed Municipal Wastewater* - A Guidance Manual. Chelsea, Michigan Lewis Publishers.

Pratt, P.F. (1972) Quality criteria for trace elements in irrigation waters. California Agricultural Exteriment Station. Page- 46.

Qadir, M., Wichelns, D., Raschid-Sally, L., McCornick, P. G., Drechsel, P., Bahri, A., and Minhas, P.A. (2010). The Challenges of Wastewater Irrigation in Developing Countries. *Agriculture Water Management* [87]: 2–22.

RCJY (Royal Commission of Jubail and Yanbu). (2010). *Royal commission environmental regulations-2010.* RCER-2010, [I-III]. Jubail: KSA Government.

Rezania, S., Din, M. F. M., Ponraj, M., Sairan, F. M. and Binti Kamaruddin, S. F. (2013). Nutrient uptake and wastewater purification with water hyacinth and its effect on plant growth in batch system. *Environment Treatment Tech*nology, [1]: 81–85.

Roberts, A. and W, Bidak. (1994). Environmentally sound agriculture through reuse and reclamation of municipal wastewater. Pp. 415-422 in *Proceedings of the Second Conference on Environmentally Sound Agriculture.* American Society of Agriculture Engineers, St. Joseph, MI.

Rudolfs, W. and Gehm, H.M. (1942). Chemical Composition of Sewage Sludges, with Particular Reference to Their Phosphoric Acid Contents. Bulletin 699. New Jersey Agricultural Experiment Station. New Brunswick, NJ: Rutgers University.

Saline Water Conversion Corporation. (2010, 2012) *Annual report for operation and maintenance.* Riyadh, KSA: Saline Water Conservation Corporation.

Saudi Geological Survey. (2012). Kingdom of Saudi Arabia numbers and facts. *Saudi Geographical Survey*, [1] Riyadh: KSA Government.

Senesi, (1989). Composted materials as organic fertilizer. *Science of Total Environ*ment, [81-82]: 521-542.

Shen, D.S., Huang, B.C., Feng, H.J., Zhao, B., Zhao, J.M., Zhang, H.Y. and Liu, P. Q. (2013). Performance of a novel decentralised sewage treatment reactor. *Journal Chemistry*, [2013]:1–6.

Sommers, L. E. (1977). Chemical composition of sewage sludges and analysis of their potential use as fertilizers. *Journal Environmental Qual*ity, [6]: 225-232.

Toze, S. (2006). Reuse of effluent water. Benefits and risks. *Agric Water Manag*, 80, 147–159.

Tzanakakis, V., Koo-Oshima, S., Haddad, M., Apostolidis, N. and Angelakis, A. (2014). The history of land application and hydroponic systems for wastewater treatment and reuse. In *Evolution of Sanitation and Wastewater Technologies through the Centuries.* IWA Publishing, London, UK, 457.

UN Department of Technical Cooperation for Development. (1985). The use of non-conventional water resources in developing countries. Natural Water Resources Series No. 14. United Nations. DTCD, New York.

Westcot, D. W., and Ayers, R.S. (1985). *Irrigation water quality criteria in irrigation with reclaimed municipal wastewater*, *A Guidance Manual*, G. Pettygrove and T. Asano, eds. Chelsea, Michigan: Lewis Publishers.

WHO. (1989) Health guidelines for the use of wastewater in agriculture and aquaculture. Technical Report No. 778. WHO, Geneva. Page- 74.

WHO. (2006). Guidelines for the Safe Use of Wastewater. Excreta and Greywater in Agriculture. [2] *Wastewater Use in Agricultur*e. WHO Press, Geneve, Switzerland.

Wintgens, T., Bixio, D., Thoeye, C., Jeffrey, P., Hochstrat, R. and Melin, T. (2006). *Integrated Concepts for Reuse of Upgraded Wastewater*; AQUAREC: Aachen, Germany.

In: Wastewater Treatment
Editor: Adriana Magdalena

ISBN: 978-1-53616-370-4

Chapter 2

PERFORMANCE OF MUNICIPAL WASTEWATER TREATMENT PROCEDURES

***Ibrahim Chourouk*[1,2,*], *Hammami Salah*[3] *and Hassen Abdennaceur*[1]**

[1]Centre of Research and Water Technologies (CERTE), Laboratory of Wastewater Treatment, Techno Park of Borj-Cédria, Tunisia

[2]Faculty of Mathematical, Physical and Natural Sciences of Tunis, University of Tunis El Manar, Tunis, Tunisia

[3]National School of Veterinary Medicine at Sidi-Thabet, Tunis, IRESA, University of Manouba, Tunisia

ABSTRACT

Wastewater biological treatment is one of the most common kinds of treatments used to improve the physico-chemical and microbiological quality of wastewater in the world for recycling, agricultural reuse and safe release in the natural receiving environments; mainly in developing countries and in small towns. The natural oxidizing pond system and the activated sludge procedure are the oldest and the most widely biological

* Corresponding Author's E-mail: ibrahimchourouk@yahoo.fr.

wastewater treatment ones in the world intended for the sewage treatment. However, the implementation of the rotating biological disks and the trickling filter procedures were recently carried out in some wastewater treatment plants (WWTPs). The objective of this chapter is to describe the main characteristics of these procedures and to evaluate their performance in the abatement of physico-chemical, bacteriological and virological pollution. In addition, the various wastewater treatment steps, the microbial communities and the mechanisms responsible for the microorganism elimination in these procedures were reviewed.

Keywords: wastewater treatment, biological procedures, performance, Wastewater quality, wastewater recycling

1. Introduction

Through this chapter of a specialized book, we wanted and opted to provide to students, engineers and water treatment technicians, adept or not, and beginners in this vast field of water treatment, a simple enumerative document summarizing the principal processes commonly used in the majority of rich and poor countries, and at the same time in both urban and rural areas. Interpretations of the operating mechanisms are intended in general as a training tool for learning and assimilating water treatment processes, in particular wastewater treatment. The description of the operating mechanisms of the various treatment processes provides a good idea about their performance, and consequently about the quality of the resulting treated water released into receiving natural environments, mainly the sea. On the one hand, this fact leads us to speak and to raise the problem regarding the legislation, the regulation and the standards of discharges of effluents leaving the treatment plant. On the other hand, after its treatment, wastewater can be reused for multiple purposes such as recycling of treated wastewater in some industrial areas known as great consumers of water, irrigation of agricultural parcels and green spaces, artificial groundwater recharge to prevent marine intrusion for example, etc. Furthermore, various aspects, related to microbes, purifiers and pathogens, have been well developed underlining the important role of microorganisms or microbes during the different steps of wastewater treatment.

Finally and depending on the final destination of the treated wastewater, which is conditioned by the sector choice to be established, the quality criteria are not the same as for a direct discharge at the end of wastewater treatment plant. Thus, there is an obligation to set up a specific sector, such as tertiary treatment and to choose the process or processes that are most adapted to the wanted reuse and the quality of the effluent leaving the site of purification by considering essentially the low cost and the technological simplicity strategies. In light of the complexity of this field and diversification of the market, we would be cautious in foreseeing a long-term evolution. However, concerns about environmental changes and impact on human health and especially water resources would drive towards: *(i)* the development of near real-time diagnostic tool and way, we mean shorter response time, more accurate results, better monitoring and regulation of processes involved in the line of treatment, *(ii)* the profit and advancing in the direction of the molecular biology progress leading to the emergence and use of new low-cost techniques.

For all these last noticed arguments, a high amount of wastewater is released worldwide into various aquatic ecosystems. From various sources as homes, hospitals, factories, etc., these types of water could carry diversified pollutants in solution or in suspension such organic or inorganic molecules, heavy metals, nutritive salts, etc., and various microorganisms and microbes [1–4]. Without any prior treatment, these types of water are likely to disturb the balance of the natural aquatic receiving environments and cause hygienic problems, mainly surface and groundwater contamination. Some remediation is provided by soils, rivers and other water systems. However, the capacity of this self-purification is so largely exceeded that we are led to protect the quality of natural waters from any excessive degradation; thus, large amount of sewage is treated in municipal or rural wastewater treatment plants. This treatment leads essentially to the reduction of oxidizable materials, suspended solids, protozoa, fungi, bacteria and viruses [1, 2]. To reduce wastewater pollution, a large number of wastewater treatment plants (WWTPs), using new treatment procedures, aim at improving the physicochemical and the microbiological quality of the treated effluents. Thus, these treatments are essential for the reduction of the physicochemical and the microbial pollutant loads in order to obtain a treated wastewater in agreement with the wastewater discharge standards [1–7]. This book chapter describes the four main useful operating biological wastewater treatment procedures: the activated sludge, the natural lagoon, the rotating biological disk and the trickling filter. In addition, the different wastewater treatment stages, the microbial communities circulating in wastewater and the main mechanisms responsible for the microorganism removal were examined in order to assess the performance of all these procedures concerning the microbial abatement.

2. Wastewater Treatment Plants (WWTPs)

Municipal wastewater is collected by a complex sewage network and directed to the wastewater treatment plant for treatment before being discharged into the natural environment or reused for various purposes as described earlier. In the plant, steps of treatment mainly vary according to the nature of the raw wastewater to be treated (from municipal or industrial origins), the future potential reuse of treated wastewater, and the sensitivity and the susceptibility of the treated wastewater-receiving environment to pollution. Currently, wastewater treatment plants have become compact, covered, automated and deodorized. They are implementing more and more efficient wastewater treatment procedures capable of removing both organic, inorganic and microbial pollution. These sewage facilities are generally installed and dimensioned to treat a certain pollution load and to ensure a certain quality of the wastewater effluent that fulfills standards of discharges in natural receiving environments.

2.1. Main Wastewater Treatment Steps

From the time of arrival of the WWTPs to that of the discharge in receiving environment, a wastewater treatment plant generally includes the different following steps [8–9]:

2.1.1. Preliminary Treatment or Pretreatment

The effluents or raw wastewater must undergo, before the real treatment, a pretreatment that involves a certain number of physical or mechanical operations. The purpose in this primary step of pretreatment is to extract and remove, from the wastewater, the solid elements in suspension or in flotations that could constitute an inconvenience for the subsequent treatments. At the point of arrival at the plant, the wastewater from the network sewers passes between the metal bars of a grid (rough screening) that allows to retain the coarse waste particles such as papers, plastics, bottles and various coarse objects. In addition, sand and gravel will be removed by settling in the bottom of specific basins since these solid materials could damage downstream installations and further handicap the process of purification. In parallel, the injection of fine air bubbles into a pond allows the oils and greases to float on the surface where they could be easily scraped (oil and grease removal) [8].

2.1.2. Primary Treatment

The primary treatment does not guarantee a release of high quality in a natural receiving environment. However, this primary treatment removes more than half of suspended solids and constitutes a non-negligible pre-purification step. This process uses different physical or chemical techniques. A primary decantation removes around 90% of suspended inorganic and organic matters that are deposited at the bottom of specific basins, called Clarifiers. All these deposited materials constitute the primary sewage sludge that is recovered by scraping from the bottom of the basins [8, 9]. On average, the pathogen removal rates with conventional primary treatment are, on average, about 1 log unit [9].

2.1.3. Biological Treatment

The biological process or treatment represents in general a secondary treatment. This secondary treatment is based on the biological purification procedure. During this treatment, most of the biodegradable pollutants are eliminated. After the primary settling or first clarification, the effluent is called the primary wastewater. Afterwards, this primary wastewater will be introduced into a specific basin equipped with aeration devices where microorganisms and various microbes, naturally present in the effluent, degrade the solidified and especially dissolved organic matter. The blown air provides for microorganisms including microbes the needed oxygen, which conditioned and improved their growth and their consumption of the organic and inorganic pollutants. This microbial growth and pollution consumption allowed exertion of physical effects on the pollutant retention by the propensity of the microorganisms to gather in flakes or films. This phenomenon is called or known as biofilm formation or production. This phenomenon is very important. Indeed, the production of microbial biofilms fundamentally contribute to the pollution abatement of sewage by increasing and improving directly the process of sedimentation or clarification since there will be a transformation of the majority of dissolved compounds on particle compounds as biofilms. In addition, the microbial biofilm formation could lead to the protection of some pathogenic microorganisms. This pathogenic organism protection is very underlined and documented in the literature.

The secondary wastewater clarification essentially allows to separate, by decantation, the treated sewage from the secondary sludge. Thus, a portion of the secondary sludge is evacuated to the sludge treatment. The remaining of this sludge is brought back to the aerated basins to maintain the biological mass needed for the seeding of the installation operation [8, 9].

The recirculation and extraction of sludge rates are calculated as described below.

The recirculation flow rate is:

$$Qr = r \times Q \text{ average}$$

With r: recirculation rate.

The recirculation rate is defined by r = Sb/(C-Sb)

Where C is the concentration of sludge at the bottom of the clarifier.

It is assumed that C = I/1000 with Mohlman index. In practice, we admit I = 150.

The Mohlman Index (MI) is the Sludge Settlement Index. This index defines (in milliliters) the volume of activated sludge decanted in 1/2 hour compared to the mass of dry residue of this sludge (in grams of material).

Therefore, we have: C = 1000/150 = 6.66 g/l

r = 4/(6.66-4) = 1.5

The recirculation flow rate is therefore

$Qr = 1.5 * 450 = 675\ m^3/h$

The sludge extraction rate, Qe, is defined by:

Qe = mass of sludge to be removed per day/concentration of sludge.

It comes so: Qe = 2242/4 = 560.5 m^3/day (or 2242 kg DM/day). With DM: Dry matter.

In conclusion, the removal rates of the main physico-chemical parameters were as follows: 95% for BOD_5, 87% for COD, 92% for SS and 65% for $N\text{-}NH_4$. In addition, the microbial organism removal rate of secondary or biological treatment is in general around 2 log units.

2.1.4. Tertiary Treatment

To further purify the wastewater and remove all the suspended solids and microorganisms, it may be necessary to carry out the following additional treatments [7, 9–11]:

2.1.4.1. Nitrogen Removal

In most wastewater, the nitrogen is in the organic or the ammoniac form (NH_4^+). A correct oxygenation in the aeration basins allows bacteria, by a nitrification process, to transform organic nitrogen into ammonia, and then to oxidize ammonia to nitrate (NO_3^-). This phenomenon of oxidation is commonly known as nitrification. Later, the nitrogen as nitrates will be transformed into gaseous nitrogen N_2 in anoxic and deficient oxygen conditions, so the absence of dissolved oxygen and the presence of oxygen combined with nitrates activated and operated as denitrification. Thus, "denitrifying bacteria" essentially reduce the nitrate ion NO_3, easily soluble in the water through hydrogen bonding with the water molecules, generating successively nitrite ions NO_2^-, nitric oxide NO, N_2O (nitrous oxide) and finally in nitrogen N_2. This N2 will be released into the atmosphere.

Thus, it is necessary to intermittently stop the aeration or oxygenation of the water when carrying out this step called denitrification, release and a loss of nitrogen from the wastewater. It should be noted that in many facilities this step of denitrification is not distinct from the other secondary steps of treatment since it is performed at a low load in the sludge basin. It will be sufficient to alternate the phases of aeration and anoxia.

2.1.4.2. Phosphorus Removal

The most widely used technique for the purification of phosphorus is the chemical precipitation by the addition of metal salts (iron or aluminum), or lime. Phosphates precipitated as metal salts or hydroxide and will be separated and removed from the liquid phase by decantation during clarification. The main reagents, which could be added, are alumina sulfate, aluminate of soda, ferrous sulfate, ferric chromium, ferric chlorosulfate and lime. The addition of these kinds of reagents could be done after the pre-treatments (the precipitation), or at the downstream of the clarifier, on the purified effluent (post-precipitation), and/or directly on the aeration basin (the simultaneous precipitation). This process is commonly known as coagulation-flocculation materialized by forming agglomerations of particles (flocs) that are more amenable to solid—liquid separation. This process of coagulation-flocculation is used in conjunction with settling or decantation, sand filtration and centrifugation to improve the efficiency and the quality of water during the later other steps of treatment. This coagulation-flocculation is an ancient technology and it is engaged as a crucial and part of modern water treatment for a century, specifically for loaded industrial wastewater and drinking water as a basic refining treatment.

2.1.4.3. Microorganism Removal

The purified wastewater contains over a million microorganisms per liter, some of which are very harmful to humans. The reduction of the number of microorganisms is essential when the treated water is discharged into a catchment area for the supply of drinking water or bathing. This microbial reduction is mainly made on a sand filter that retains the last suspended particles and attached microorganisms or by chemical disinfection [7]. Several types of treatment procedures were used for wastewater disinfection, such as chlorination, ultraviolet irradiation, ozonation and peracetic acid [7, 9]. Chlorination approach is the most commonly used disinfection method in sewage treatment plants for microorganism inactivation in wastewater before their release into the receiving environment or for the agriculture reuse [12]. This method, which has a lower cost compared to other disinfection techniques (ultraviolet irradiation and ozonation) [7, 10], is effective for microorganism removal, and it is easily used in wastewater treatment plants. In addition, chlorination has economic, financial, environmental and public health benefits [13]. However, the by-products and the residues formed during the wastewater chlorination such chloramines, the trihalomethanes and chloroform [13], are known to be toxic and persistent in the environment. Disinfection of wastewater by UV radiation is proposed as an alternative solution to the chemical disinfection that showed various harmful effects to man and his environment. The ultraviolet irradiation is achieved by the conversion of electrical energy in non-ionizing electromagnetic radiation. The most efficient wavelength for the inactivation of microorganisms is around of 260 nm [14]. The high photon energy radiation is intensively absorbed by the genetic material, RNA or DNA. The absorbed photon energy modifies certain chemical bonds, especially at the level of thymine nucleotide bases and generating thymine bridges. This structural perturbation leads to the cessation and interruption of duplication and replication processes of the genetic material. Thus, the direct consequence of UV exposure is the inability of the cell to reproduce and viral particles to infect correctly specific host cells, or the appearance of some mutants that could no longer ensure all their vital functions [7, 11]. This mode of disinfection is also characterized by a short contact time and a more effective virucidal action. The specific energy required for total or partial destruction of microorganisms depends on the physicochemical composition of the wastewater to be treated, and the sensitivity of the microbes [7, 11]. Thus, microorganisms are ranked in ascending order of resistance as follows: bacteria, viruses, fungi, spores and cysts [11]. The only problem with UV disinfection is that some microorganisms are able to repair the damage of the genetic material caused by the UV lights, near-UV and/or violet-blue after exposure to the day light. This phenomenon is known as photo reactivation [11]. On average, the removal efficiency of enteric viruses and pathogenic bacteria of tertiary treatment with UV-C is around 3 (99.9%) to 4 (99.99%) logs 10 units [9].

2.2. Biological Wastewater Treatment Procedures

The common biological purification procedures are multiple and of various designs. They are essentially based on the ability of microorganisms to assimilate polluting materials. The principle is similar to the one existing in natural environments and exerted by various autochthon purifiers. Therefore, the main biological treatment procedures are as follows: intensive procedures with free cultures (activated sludge), fixed culture procedures (trickling filters, rotating biological disks or rotating biodisks) and extensive procedures (natural oxidizing lagoons) [8].

2.2.1 Activated Sludge Procedure

The main method used for the purification of polluted effluents is activated sludge procedures due to its low cost and its high effectiveness [8, 15]. This method is more effective for the organic pollution removal in municipal and domestic wastewater than the other biological wastewater treatment procedures, such as natural lagoons, rotating biodisks and trickling filters [15–17]. This activated sludge procedure is still called the free culture system because the purifying microorganisms float and circulate freely in the effluent [18]. It reproduces the action of a river with an intense dynamic of aerobic microbial populations maintained in suspension. In this compartment, stirring and oxygenation devices, continuous or intermittent, are essential for the activity of the purifying microorganisms [18]. It consists of placing the wastewater in contact with a purifying biomass in this specific aerated and turbulent environment. Thus, all small organisms represented by various microflora or microbiotas, microfauna of animals and protozoa, acted as biological operators for consummation and degradation of the main complex pollutant compounds existing in wastewater. The degradation is then achieved essentially by aerobic pathways, in the presence of oxygen and by the transformation of the various pollutants as microbial biomass [8, 16]. The intensive growth of aerobic microbes follows the high availability of organic matter provided in the water and the presence of appropriate amount of dissolved oxygen. At the end of this important process, generated microbial biomass settle in the water basin digester. To enhance this purification phenomenon, a free bacterial culture as developed flocs will be supplied with water to be purified in broad and ventilated basins. An agitation is carried out on the surface by means of a turbine, or at the bottom of the basin by diffusion of air bubbles. This action aims at avoiding deposits and homogenizing of the suspension mixture of bacterial flocs and wastewater. The aeration or air oxygen diffusion is intended to enhance dissolved oxygen production in the water, and thus to meet the needs of aerobic purifying bacteria. The contact time between wastewater and biomass varies between 6 to 10 hours depending on the season and the quality of raw wastewater. After this time, the purified effluent is separated from the microorganisms by secondary decantation. A portion of the formed sludge will be recycled in the aeration basins as earlier evoked to ensure the reseeding in a microorganism and the sludge excess being extracted and treated [8].

2.2.2. Natural Oxidizing Lagoon Procedure

Natural lagoons or natural oxidizing lagoon procedure is the oldest and the most frequently used biological wastewater treatment procedure for municipal wastewater purification [7, 12, 19]. Unlike other methods, natural oxidizing lagoons and derived techniques are characterized by slow microbial kinetics [10]. However, these methods have undeniable advantages such as a low cost of installation and operation, good removal of the microbial load and low production of sludge. They are most frequently used in many countries. The lagooning process is a natural treatment technique that consists in the accumulation of wastewater in ponds, lagoons or basins, noted as biological or stabilization ponds, where a series of complex biological, biochemical and physical progressions and developments take place. Lagoon is a stationary system showing a continuous flow: several ponds working in parallel in which the inlet flow and the outlet flow are equal form lagoon plants. The pond depth ranges from 1 (aerobic ponds) to 5 m (anaerobic ponds). Wastewater stays in the line of ponds from 10 to 30 days in order to be treated. Thus, the residence time varied according to the dimension of the different ponds, their depth in particular, and the quality of raw water to be treated. Lagooning is known as a very efficient process for wastewater treatment method, and commonly implemented at both the scale of small rural communities or at medium or large urban ones. In certain specific cases, lagoon plants could be built to treat wastewater released by food seasonal industries.

Thus, the lagooning or stabilization ponds process consists of retaining pretreated effluents in more or less deep basins exposed to the atmospheric air for periods of up to several weeks. During this time, the biodegradation and the removal of the pollutant load take place. Lagooning is a natural procedure of sewage treatment that allows a separation of the solid elements of the liquid phase by sedimentation. It is a biological purification mainly due to the action of bacteria and algae. It is also known as an extensive wastewater treatment procedure where the water to be treated circulates slowly in large shallow basins or lagoons. These basins could be subdivided into three geometric zones: a superficial ventilated zone, a medium optional zone and a deep anaerobic zone. The purification of wastewater is achieved by natural mechanisms similar to those observed in self-purification of rivers with algae and ordinary bacteria and without control of the concentration of active biomass. It is a mixture of aerobic, anaerobic and facultative microorganisms [1, 2, 10]. The operating principle of stabilization ponds rely on the settling of suspended solids in shallow basins with high residence times in which living microorganisms proliferate to the detriment of the organic matter contained in these waters [1, 2, 10, 20]. In this type of process, the two aerobic and anaerobic purification procedures coexist. Wastewater circulates in a series of successive basins (at least three) and they are retained in the first anaerobic basin located in the upstream of the process. At the water-sediment interface, anaerobic bacteria stabilize the decanted organic matter that is transformed into methane (CH_4), carbon dioxide (CO_2) and in ammonium. In aerobic upper parts, the non-decanted organic matter released from the sediments is degraded by aerobic bacteria and this degradation is promoted by the oxygen provided by algal photosynthesis. Thus, each type of stabilization ponds forms a distinct ecosystem [1–4]. The balance between the different stages of degradation of organic matter could be severed by many intrinsic and extrinsic parameters. So, climatic factors (mainly average temperature, sunshine and wind speed), wastewater characteristics (inorganic and organic loads), layer depth of sediments accumulated in the bottom of each basin, the lagoon design (length, wide and especially the depth) all act on the microbial species composition and their abundance [3], and therefore directly on the performance of the system.

2.2.3. Rotating Biological Disk Procedure

Rotating biological disks or rotating biodisks are an aerobic biomass treatment procedure with fixed biomass. The supports of the purifying microflora are disks partially immersed in the effluent to be treated and driven by a rotational movement which ensures both mixing and aeration [2, 10, 21]. The microorganisms grow and form a purifying biological film on the surface of the disks. The disks are partially immersed; their rotation allows the oxygenation of the fixed biomass. The effluent is previously decanted to avoid the clogging of the support material. The sludge that falls out is separated from the treated water by clarification. The biological disk unit consists of rotating plastic disks mounted on a shaft in an open basin filled with wastewater. Disks rotate slowly in the pond and as they pass through the wastewater, organic matter is absorbed by the biofilm attached to the rotating disk [1, 2, 10, 22]. The accumulation of biological materials on the disk increases the thickness and forms a deep layer of sludge. This sludge structured as a thick layer of microbial biofilms represents essentially the support for the purifier organisms and it will be alternatively exposed to air or immersed in water with the biodisks. Thus, when the biodisks are exposed to air, oxygen is absorbed and promotes the growth of this biomass organized as a thick layer of microbial biofilms. When the microbial biofilm is too thick, enough amounts come off and settle in the bottom of the unit. Furthermore, the alternation process of contact phases with the air (oxygenation of the biofilms) and the effluent to be treated (immersion of the biofilms), following the rotation of the biodisks support, allows the oxygenation of the system and the development of the bacterial culture. During the immersed phase, the biomass absorbs the organic matter that will be assimilated by aerobic fermentation conditioned by the atmospheric oxygen of the emerging phase [1, 2, 10, 22].

2.2.4. Trickling Filter Procedure

Trickling filter is an aerobic biological wastewater treatment method with fixed culture [10, 21, 23]. The microorganisms develop on a support material regularly irrigated by the effluent to be treated. This procedure consists of supplying water, previously decanted and a structure containing a mass of material (pozzolana or plastic) serving as support to purifying microorganisms that form a biological film responsible for the pollution assimilation. The biofilm detaches gradually as water percolates. At the exit of the trickling filter, a mixture of treated water and biofilm is collected. The latter is trapped at a secondary clarifier in the form of sludge and the treated water reaches the natural environment. This purification method is based on the capacity of the microorganisms contained in the effluents to synthesize and release viscous exopolymers (long polysaccharide filaments called Glycocalix) that ensure their fixation on an inert support. The bacteria agglomerate to form a biofilm. The biofilm is essentially composed of bacteria, but also hosts other organisms integrated in a more or less complex food chain (protozoa, metazoans, insects…). The biofilm is self-regulated by the natural detachment. The absence of oxygen in the bottom of the floc causes anaerobic fermentation. This fermentation induces the formation of microbubbles and a weakening of the biofilm. This biofilm mainly detaches and will be removed with the effluent, and equally be consumed in majority by some macro invertebrates (nematodes and insect larvae) and different species of protozoa (amoebae, flagellates). The operating principle of a bacterial bed consists of trickling the treated water, on a specific large surface area material, serving as a support for microorganisms and forming a more or less thick film. The aeration is practiced most often by natural draft and sometimes by forced ventilation. This aeration is intended to provide, throughout the mass of the bed, the oxygen necessary to maintain an aerobic activity of the microflora. The pollution and the dissolved oxygen contained in water diffuse through the biological film up to the microorganism assimilators, while conversely the reaction products are eliminated in the circulating liquid and gaseous fluids. The purifying performances are therefore directly related to the activity of anaerobic biomass conditioned by the hydrodynamics of the process and the transfer of material between phases, in particular the transfer of oxygen [10, 21, 23].

2.3. Microbial Communities in Wastewater Treatment Plants

Several groups of prokaryotic and eukaryotic organisms find favorable conditions for their proliferation in sewage treatment plants. These are the scrubbers of polluted effluents. These multiple species may vary depending on the treated effluents, climatic conditions, organic load and water depth. The main groups are bacteria, algae and zooplankton [7].

2.3.1. Bacteria

In all wastewater treatment procedures, bacteria represent the largest proportion of microbial biomass and play the crucial role in purification [7]. Bacteria can degrade and assimilate a lot of the organic matters contained in wastewater. These bacteria release into the environment degradation products that are essentially soluble mineral matter and dissolved gases. According to the balance of the environment, in particular nitrogen and phosphorus rates, the most suitable bacteria grow rapidly and dominate the other species. There is a natural regulation of the bacterial growth rate, in relation with the organic matter present in wastewater and other growth conditions, such as temperature, sunshine, pH, oxygen, dissolved oxygen. Two different types of bacteria are found in biological wastewater treatment procedures. The aerobic bacteria, which, in the presence of dissolved oxygen, transform the dissolved organic compounds to inorganic matter (as nutrients) and gases. The anaerobic bacteria that are essential methanogenic (methane formation) perform the transformation of organic matter in sludge to methane. In an activated sludge procedure, the density of bacteria reaches 10^9 cells/mL [6, 24]. These bacteria associate with the dissolved or colloidal molecules and with the particles of the medium to form insoluble elements or flora. The bacterial flora of the natural lagoons and the activated sludge procedures are generally similar and the dominant genera are Gram-negative [6]. In addition, the bacterial flora is subdivided into two large groups: unicellular bacteria and filamentous bacteria known as actinobacteria or Actinomyces.

2.3.2. Zooplankton

The second-largest group of microorganisms in wastewater treatment plants is zooplankton. The wildlife has an essential importance in the operation of lagoons and many organisms actively participate in the purification of the environment (predation, antibiotic activity, filtration, etc. [7]. The protozoa are an important and main predatory of bacteria. They constitute the only winter abundant zooplankton in the last lagoon ponds. This function has been shown in the case of activated sludge. The protozoan ciliates are indicators of the quality of the effluent and the proper functioning of an installation [25]. The rotifers are microscopic vermicides (Vermidiens); they actively filter phytoplankton and are able to survive with very low dissolved oxygen levels. The Cladocera's are also small crustaceans that favor the abatement rate of suspended solids. Thus, they allow a lightening of the medium and the penetration of light. On the other hand, they cause a decrease in the dissolved oxygen level because of their breathing and the microalgae elimination.

2.3.3. Algae

Algae are microscopic planktonic organisms. They are mainly represented in the lagoons by the following species: blue algae (Cyanophyceae) close to the bacteria, green algae (Chlorophyceae), brown algae (Charophyceae) and the euglenas [7]. The abundance and diversity of microalgae depend on the season and charges in nutritious substances [6]. The algae grow in the light by taking carbon dioxide and mineral salts and eliminating the oxygen necessary for other aerobic organisms in the environment. Algae are the main oxygen producers in the lagoons. This production is carried out largely in the surface water layer (up to 40–50 cm). Thus, algae contribute to the reduction of nutrient salts introduced into the pre-treated effluent.

2.3.4. Enteric Viruses

Some other organisms, not associated with purification, are found in wastewater. They are vectors favoring many germs and parasites often of fecal origin, such as pathogenic bacteria [1, 22, 26], enteric viruses [1–4], parasitic protozoa [7]. Viruses are found in abundance in the water environments [1–4]. They are excreted up to 10^{12} per gram of feces [27], and they are extremely infectious [28]. The viruses of the digestive tract, called enteric viruses, are free in feces, and they are conveyed by the sewage. These viruses excreted in the feces are all non-enveloped and very resistant in the environment and their mode of transmission is completed essentially through the fecal-oral pathway. Large numbers of these viruses are excreted in the feces of patients or healthy carriers. According to the associated pathologies ranging from simple viral gastroenteritis to severe hepatitis, three groups of enteric viruses are distinguished: gastroenteritis viruses (Caliciviruses, Rotaviruses, Aichi viruses, Astroviruses and enteric adenoviruses 40/41) and the enteroviruses (Polioviruses, Echoviruses, Coxsackie viruses) and hepatitis viruses (Hepatitis A and E). The quantities of infectious particles released by the infected subjects are considerable and can reach 10^9 per gram of feces [29]. In addition, the average duration of viral excretion for hepatitis E viruses (HEV) is one month, but often it persists for several months [30]. It starts 5 to 6 days before the onset of clinical signs and reaches between 10^3 and 10^6 particles per gram of stool. However, the adenovirus particles are excreted for 7 to 14 days at a rate of 10^6 to 10^7 per gram of feces [31]. The Noroviruses appear in the feces with the first symptoms and they are present for 3 days at a rate of 10^6 per gram of stool. The arboviruses are excreted at the same rate over a period of 12 days [31]. The rotaviruses are excreted for 10 days; their elimination is maximal (10^{12} viral particles per gram of feces) during the first 3 to 4 days of infection [32]. The transmission of these viruses to humans generates several diseases that affect all age groups. Enteric viruses are generally present in natural water and wastewater. They are the main cause of waterborne and foodborne outbreaks following the consumption of water or food contaminated by fecal microorganisms [1–4, 33–35]. Viral gastroenteritis is the second most common cause of morbidity after respiratory infections [36]. Diarrhea remains the second-leading cause of death, mainly in developing countries, among children under 5 years of age, causing approximately 1.3 million preventable deaths each year, [37]. Once, these viruses are excreted in feces at relatively high concentrations, they will be present in the wastewater where they are only partially removed by the wastewater treatment procedures. The survival and the resistance of these viruses to the treatment procedures facilitate their diffusion and transmission through the fecal-oral route. This transmission can be directly from person to person or from animal to human or indirectly through the ingestion of food or contaminated waters [1–4, 33, 38]. Several potential routes ensure viral transmission to humans, such as the use of domestic surface water or contaminated groundwater, swimming in dirty water, consumption of infected shellfish and crops grown in contaminated soils. The human is considered the primary contaminant and the secondary receiver [7].

2.3.5. Minimal Infecting Dose (MID)

For infectious organisms, dose-response relationships have long been described by point values such as the Minimal Infective Dose (MID). This corresponds to the smallest number of pathogens that must be absorbed for symptoms of the disease to occur in at least a few subjects. It establishes a relationship between the level of exposure to microorganisms and the probability of occurrence of developing a deleterious effect. The literature does not provide minimum infective dose values for all existing microorganisms. In most cases, orders of magnitude are available for different families of pathogens. In general, these data showed the infectivity of viruses and protozoa 10 to 100 times higher than bacteria.

2.4. Virus Survival and Factors of Viral Removal in Wastewater Treatment Plants

The survival of enteric viruses in sewage treatment plants is often related to a phenomenon of aggregation that is generally described as essentially dependent on the association of viruses with solid particles. Particulate elements in suspension, the sludge and the sediments protect and safeguard viruses against inactivating agents, such as heat and disinfectants (chlorine and ultraviolet radiation) [10, 11]. In the wastewater treatment plants, the elimination of viruses is carried out only by natural mechanisms, such as the adsorption of viruses to solids, the predation of viral particles by protozoa and other natural inactivation factors related to direct and indirect sunlight [7, 19]. In addition, the main factors that can influence the survival of viruses in wastewater are temperature, sunlight, cation presence and biological factors [7].

2.4.1. Virus Adsorption to Solid Particles

In wastewater, the fate of viruses depends on a number of parameters. These parameters are difficult to segregate because of their interference with each other. The situation of viruses in the wastewater is particularly complex because the enteric viruses are rarely free and isolated. They are present in the wastewater most of the time in aggregated form or associated with either organic matter or particles in suspension. As a result, these viruses are more or less protected against inactivation by some physico-chemical and biological factors present in natural environments [38]. Indeed, high temperature and extreme pH play an important role in viral inactivation by destruction of their nucleic acids and proteins. However, in wastewater treatment plant viruses' behavior is extremely different from that of isolated viruses, especially with regard to sensitivity to oxidizing agents. The pH and temperature effects are of great impact on their interaction with other environmental factors [7, 41]. Usually, viruses are inactivated at alkaline pH (> 11) during treatments with lime or ammonia, which have a direct toxic effect on viruses by fragmentation of nucleic acids [42]. However, these viruses resist in general much better at acid pH, such as pH 3. The natural inactivation of enteric viruses is also very dependent on the temperature. High temperature facilitates the inactivation of viral particles. Kott et al. [43] reported poliovirus inactivation within 70 days at 18–23 °C in oxidant basin effluents. Virus survival is also affected by the presence of organic or inorganic ions in the water, such as the thermostabilizing of naked RNA viruses in the presence of divalent and trivalent cations [44]. In addition, di and trivalent ions indirectly protect viruses by promoting viral adsorption on clays. The mechanisms involved in the phenomenon of virus adsorption to solid particles are those related to the theory principles of extended Derjaguine Landaue Verweye Overbeek (DLVO). The last theory based on the action of the Van der Waals forces, electrostatic double layer forces, hydrophobic and osmotic interactions between particles [7, 45]. Indeed, the electrostatic interactions play an important role in the association of particles with viruses. This type of interaction is influenced by the electrical charge of the capsid or the viral particle surfaces [46]. This charge is generated after the ionization of the different carboxyl groups and amines situated in a virus capsid at various pH levels [47]. In addition, the hydrophobic interaction may be explained by the hydrophobicity degree of viral capsid proteins that promote the aggregation of viruses to solid particles at pH levels above their isoelectric point [48]. All these types of interactions are dependent on the virus characteristics, the various suspended particles and the kind of water [7].

2.4.2. Viral Particles Predation by Organisms of Higher Trophic Levels

The influence of microbial biomass is manifested in different ways. The ingestion of viruses by protists in water environments has recently been reported in the marine domain by various algae (Chlorella) and bacteria in sewage, surface or seawater (Vibrio, Flavobacterium, Klebsiella, Pseudomonas…). These kinds of microorganisms have also been suspected of having some antiviral action. Because of the different experimental conditions, it is difficult to compare the results and accurately determine which bacterial species are causing the viral inactivation in the natural environment. However, Girones et al. [49] isolated marine bacteria of the genus *Moraxella* possessing specific antiviral properties against polioviruses. Similarly, Ward et al. [50] isolated from surface water 27 bacterial strains that showed virucidal effect, in particular against Enteroviruses.

The influence of the microbial biomass is expressed in different ways. The ingestion of viruses by protists in water environments has recently been reported in the marine domain. The viral particle predation by protozoa is now admitted by the researchers [51]. Under laboratory conditions, *Tetrahymena pyriformis*, taken as a model for ciliated Protozoa, is capable of phagocyting simian rotavirus SA 11 particles. Initially, placed in medium with the protozoa, the viral particles were found 90 min later in the digestive vacuoles of the cilia cell [52]. Several studies reported the internalization of enteric viruses by protozoa in *vivo* and in *vitro* [53], protecting them and ensuring their replication and transmission [54].

Battistini et al. [55] described the absorption of adenoviruses by protozoa, where it can resist for 35 days. Another aspect of the influence of microorganisms is the non-specific adsorption of viral particles on prokaryotic or eukaryotic organisms. A decrease in viral concentration of 30 to 50% was obtained after the contact of poliovirus 1 with pure or mixed cultures of bacteria [56]. Similarly, many studies revealed the consumption of virus-like particles by the heterotrophic nano-flagellates [57, 58]. There are limited data in the literature regarding the enteric virus predation in WWTPs. Further research is needed to better understand the enteric virus predation by higher trophic organisms in WWTPs.

2.4.3. Sunlight

Sunlight has an influence on the survival of viruses by direct action on the viral particles, and indirectly by stimulation of the microfauna development. The viral inactivation occurs at a wavelength of less than 370 nm [59]. Hurst [60] who reported that in the absence of sunlight, the inactivation rate varied from 0.71 to 0.80 logs per day, whereas in the presence of sunlight, rate varied from 1.33 logs in very turbid water to 2.38 logs per day in water of low turbidity. However, sunlight action is decreased in the presence of suspended solids, which protect the adsorbed viral particles and limit the diffusion of ultraviolet rays. In fact, the viral inactivation is the result of a photochemical reaction that causes an alteration of the nucleic acid or other non-nucleic receptors of the virus under the effect of ultraviolet rays and visible spectrum rays [61].

Three different mechanisms of sunlight that can promote the viral inactivation in biological wastewater treatment procedures (especially the natural lagoon and the activated sludge procedures) such as the direct inactivation and the indirect endogenous and exogenous inactivation [7, 14]. The direct mechanism is more efficient in radiation with wavelengths ranging between 100–280 nm (UVC) and 280–315 nm (UVB) [14]. These two types of UV irradiation involve the uptake of photons by viral particles, causing viral genome replication blockage and/or viral capsid protein hydrolysis [7]. In fact, the type and the size of the viral genome influenced the sunlight performance concerning the virus inactivation.

Indeed, the virus absorbance of sunlight photons induced the thymine dimer alteration. Thus, the RNA viruses are more resistant to direct sunlight inactivation than DNA viruses [62]. Similarly, viruses with small genomes are more resistant than viruses with large genomes [62].

The indirect sunlight inactivation mechanism involves the reaction between the sunlight photons and sensitizer molecules, allowing the radicals or reactive species formation that inactivates the viral particles by causing the damages in viral nucleic acids or proteins [63]. These sensitizer molecules can be in the viruses themselves (endogenous inactivation) or other molecules that circulate in wastewater (exogenous inactivation) [7, 14]. This indirect mechanism is more effective in the UVA (315–400 nm) and visible radiation (400–500 nm) [7, 14]. The endogenous photosensitizers of viruses may include the amino acid chromophores situated in the viral capsid proteins [41, 64]. However, the exogenous photosensitizers may contain the alga compounds and the natural organic matter [65]. The two mechanisms of indirect sunlight inactivation (exogenous and endogenous) are more efficient in the maturation basins than the facultative anaerobic basins in biological wastewater treatment (natural lagoons and activated sludge). In the maturation basins, the dissolved oxygen rate is very high and the penetration of sunlight in these basins is more important than in other basins. Consequently, the pathogenic microorganism abatement in the maturation basins is very crucial to other basin types [7]. Many reports showed the effectiveness of the indirect sunlight inactivation mechanism for MS2 coliphages [41, 64]. However, several recent studies revealed the resistance of different types of enteric viruses such as Rotaviruses A and Adenoviruses to indirect exogenous sunlight mechanisms [41, 66]. In summer, the different types of sunlight inactivation mechanisms are related to the virus's characteristics (genome type and size, viral capsid proteins), to the solar radiation wavelengths, and to the physico-chemical and optical parameters of WWTPs [7].

3. PERFORMANCE OF BIOLOGICAL WASTEWATER TREATMENT PROCEDURE

Table 1. Purification performance of various municipal wastewater treatment procedures

Procedures		Physico-chemical parameters				Bacteriological parameters		Virological parameters
		SS (%)	COD (%)	BOD_5 (%)	$N\text{-}NH_4$ (%)	FC (U $\log_{10}$)	FS (U $\log_{10}$)	Enteric viruses (U $\log_{10}$)
Pre-treatment	Bar Screen	—	—	—	—	—	—	—
	Sand removal	-	-	-	-	-	-	-
	Oil removal	-	-	-	-	-	-	-
Primary Treatment	Primary Clarification	50	40	40	70	0 - 1	0 - 1	0 - 1
Secondary/ Biological Treatment	Activated Sludge	92	87	95	65	1.5 - 2	1.5 - 2	1 – 2
	Trickling Filter	75	65	90	80	1 - 2	1 - 2	1 - 2
	Rotating Biodisks	75	65	90	70	1 - 2	1 - 2	1
	Natural lagoons	64	60	65	52	3	3	1
Tertiary Treatment	Phosphorus removal	-	-	-	-	-	-	-
	Nitrogen removal	-	-	-	-	-	-	-
	Disinfection					3	3	2-3
	Chlorination					2	3	3-4
	Ultraviolet					2	3	3
	Ozonation					2	4	3-4

SS: Suspended Solids, **COD**: Chemical Oxygen Demand, **BOD_5**: Biological Oxygen Demand, **N-NH4**: Ammonium Nitrogen, FC: Faecal Coliforms, FS: Faecal Streptococci, **-** Not reported.

The purifying performances of the main wastewater biological treatment procedures were determined based on their effectiveness in the physico-chemical, bacteriological and virological pollution abatement. The comparison of the performance of these procedures for the physico-chemical pollution reduction is based on the criteria specifically characterizing the wastewater pollution, such as soluble chemical oxygen demand (COD), soluble biochemical oxygen demand (BOD_5), suspended solids (SS) and ammoniacal nitrogen (NH4+). Similarly, the performance assessment of these procedures for the bacteriological abatement is established by the enumeration of fecal indicator bacteria such as fecal coliforms and fecal Streptococci. In addition, the effectiveness evaluation of these procedures for the enteric virus removal is determined by the measurement of enteric viruses (Rotavirus A, Noroviruses genogroups I and GII, Aichi viruses, Astroviruses, etc.), detection rates and their viral loads in different basin types of each biological procedure (Table 1).

3.1. Activated Sludge Procedure

The activated sludge procedure provides an excellent physicochemical abatement and a moderate bacteriological pollution reduction [17, 67]. In fact, the removal rates are as high as 95% for BOD_5, 87% for COD, 92% for SS and 65% for $N\text{-}NH_4$ [17]. In addition, the fecal bacterial reduction oscillates between 1.5 and 2 logarithmic units (2 log units) [67] and the activated sludge procedure ensures an important improvement in the virological quality of the treated wastewater. Indeed, the enteric virus removal rates generally vary between 2.5 and 3 $\log_{10}$ units [68]. Several previous and recent environmental studies have shown some various enteric virus removal, such as Noroviruses, rotaviruses, Enteroviruses and Enteric adenoviruses by activated sludge procedures [68–70]. The removal rates of these enteric viruses could reach 3 $\log_{10}$ for human Adenoviruses and Polyomaviruses [68], 1.64 to 2.34 logs_{10} for Noroviruses genogroup I, 2.14 to 3.11 logs_{10} for Noroviruses genogroup II [69, 71], 1.31 to 2.15 logs_{10}. [9, 72–74]. These rates could attain for Rotaviruses A 2 to 3 $\log_{10}$ units (99.9%) for enteroviruses [70, 73] and 1.3 logs_{10} for Sapoviruses [71], respectively. In contrast, the activated sludge procedure sometimes allowed a poor virological quality of the treated effluents, with a low abatement rate. The treatment performance reported in the case of Aichi viruses and BK polyomaviruses varied between 0.76 and 1.00 logs_{10} by activated sludge procedures [71] (Table 1).

3.2. Natural Lagoons

The natural lagoon procedure ensures an average abatement of the carbon pollution of about 65% for BOD_5, 60% for COD, 64% for SS and 52% for $N\text{-}NH_4$ [1–4, 10, 20]. However, the average abatement of the fecal coliforms and Streptococci is around of 99.9% (3 $\log_{10}$ units) [1, 4, 10, 20]. Therefore, a poor physicochemical quality and an excellent bacteriological quality of the treated effluents were obtained by natural lagoon procedures, superior to log 10 units [1, 4, 10, 20]. Similarly, the improvement of the virological quality of the treated wastewater by this category of biological wastewater procedure was reported with moderate removal rates (44.9% for Enteroviruses, 58.4% for Noroviruses GII, 72% for Rotaviruses A) [1, 2, 7, 70]. Nevertheless, other recent studies revealed a high resistance of other types of enteric viruses such as Aichi viruses and Astroviruses to this lagooning biological treatment system [3, 4]. Therefore, the effectiveness of the elimination of biological parameters (bacteria and viruses) is obviously very important compared to that usually recorded for physicochemical parameters [1, 2, 10]. Besides, it is well known that the residence time in the biological wastewater treatment procedures plays a key role in the overall performance for the various physicochemical reductions and the biological parameter elimination in order to improve the quality of the treated wastewater. The residence time in the natural lagoon procedure is generally very high, with on average 30 days. Wastewater treatment by this procedure is ensured by natural mechanisms, such as the effect of solar radiation, the viral particle adsorption to solid particles, the activity of the microflora and the deficit of organic matter in the last three maturation basins that promote and facilitate the microorganism inactivation such as bacteria and enteric viruses [1–4, 6, 10]. The major disadvantage of this procedure is the algae proliferation in the lagoons that greatly affects the physicochemical quality of the treated effluents. They are responsible for the eutrophication of the natural receiving environments [10] (Table 1).

3.3. Rotating Biodisks

The rotating biological disk or rotating biodisks procedure provides generally a good carbon purification. The removal rates are around of 65% for COD, 90% for BOD_5, 75% for SS and 70% for $N\text{-}NH_4$ [1–4, 10]. However, this biological procedure ensures a low abatement of the fecal indicator bacteria. Indeed, the treatment performance of fecal coliforms and fecal Streptococci usually fluctuates between 1 and 2 $\log_{10}$ units [1–4, 10]. Moreover, a recent environmental report revealed the ineffectiveness of this procedure for the *Salmonella* removal [22]. Similarly, this procedure sometimes shows a low to moderate improvement of the virological quality of the treated effluents. In a recent study [2], the elimination rate of Rotaviruses A was about 50%. In addition, the other previous environmental study revealed the Noroviruses GII detection with a high frequency (84%) and a low viral load (1.5 genome copies/μL) at the exit of the rotating biodisks [1]. Furthermore, two other recent environmental studies described the high resistance and the persistence of other enteric viruses, such as Astroviruses and Aichi viruses to this biological wastewater treatment procedure [3, 4]. Therefore, the rotating biodisks procedure usually ensures a satisfactory physicochemical purification and a poor bacteriological and virological treatment. These data were explained by the residence time, which is very short (4 to 5 hours on average) in this type of biological procedure. In reality, it would be a biofiltration and the physicochemical quality of the wastewater is considerably improved compared to the one generally recorded in the case of the lagoon system or activated sludge procedures [1, 2, 10]. It is well known that the treatment with rotating biodisks affects more specifically the wastewater clarification, as a consequence the reduction of turbidity and of suspended solids [1–4, 10] (Table 1).

3.4. Trickling Filter

The trickling filter procedure provides an average reduction of 65% for the soluble COD, 90% for the soluble BOD_5, 75% for the SS and 80% for the ammoniacal nitrogen [10, 21, 23.75] However, this procedure allows low bacteriological abatements. The removal rates of the fecal coliforms and the fecal streptococci oscillated on average between 1 and 2 logs $_{10}$ units [10]. Generally, the trickling filter procedure presents an excellent wastewater physicochemical quality and a bad bacteriological one for the treated effluents, as the case of the rotating biodisks procedure [10]. Nevertheless, the wastewater purification could be achieved at a much shorter residence time (12 min), than that obtained by the rotating biodisks (3 to 4 h) [10, 21]. In addition, the trickling filter procedure also ensures a moderate improvement in the virological quality of the treated wastewater, with various log $_{10}$ reduction [71]. In Fact, there are limited published data on enteric virus elimination by trickling filter procedures. The first study reported [70] the effectiveness of trickling filter procedure for the enterovirus removal with a high abatement rate of around 95.7% (2 log $_{10}$ units). A second study [71] showed that the average log $_{10}$ reduction of enteric viruses by trickling filter was around 2.85 for Noroviruses genogroup II, 2.57 for Noroviruses genogroup I, 2.56 for JC Polyomaviruses, 2.3 for Enteroviruses, 1.5 for Sapoviruses and for BK Polyomaviruses, 1.3 for enteric adenoviruses, 1.2 for Rotaviruses A, 0.99 for Aichi viruses and 0.76 for pepper mild mottle virus (Table 1).

CONCLUSION

This chapter presents an overview concerning the performances of the main wastewater treatment procedures, such as the natural lagoons or ponds, the rotating biodisks, the trickling filter and the activated sludge for the physicochemical, bacteriological and virological pollution removal. The natural lagoon procedure provides a poor physicochemical purification, a good microbiological quality and a moderate improvement of the virological quality of the treated effluents. The algae proliferation in the lagoons greatly influences the physicochemical quality of the effluent, which is responsible for eutrophication phenomena into main natural water receiving environments. The activated sludge procedure ensures a better physicochemical purification than the one obtained by natural lagoon procedures. Moreover, this procedure assures a moderate improvement of the bacteriological and the virological pollution abatement as compared to the natural lagoon procedures. The rotating biodisks and the trickling filter procedures allow a satisfactory physicochemical purification. Nevertheless, an important inefficiency is observed at the level of the pathogens abatement (bacteria and viruses) by these two biological wastewater treatment procedures. Generally, the bacteriological and virological quality of the treated effluents by the biological wastewater treatment procedures are not in conformity with wastewater discharge standards since they else present a high viral frequency and contain a large number of enteric bacteria and viruses. However, the moderate or the bad efficiency of natural lagoons and activated sludge procedures for the fecal bacteria and the enteric virus abatement, did not allow in general the discharge of a good sanitary quality of the treated effluents for later agricultural reuse or release into receiving natural environments. Consequently, the tertiary disinfection treatment by using chlorination or UV radiation is needed and essential for the pathogenic microorganism inactivation.

ACKNOWLEDGMENTS

The authors gratefully acknowledge the support from the LTVRH staff (CERTE, Tunisia) for their technical support and, for editorial assistance during the redaction of this special book chapter.

CONFLICT OF INTEREST

The authors have declared no conflict of interest.

REFERENCES

[1] Ibrahim C, Chérif N, Hammami S, Pothier P, Hassen A. Quantification and molecular characterization of Norovirus after two wastewater treatment procedures. *Water Air Soil Pollution.* 2015; 226 (6): 187–200. DOI: 101,007/s11270-015-2402-x.

[2] Ibrahim C, Chérif N, Hammami S, Pothier P, Hassen A. Quantification and genotyping of rotavirus A within two wastewater treatment processes. *Clean Soil, Air Water.* 2016; 44 (4): 393–401. DOI: 101,002/clen.201400588.

[3] Ibrahim C, Hammami S, Mejri S, Mehri I, Pothier P, Hassen A. Detection of Aichi virus genotype B in two lines of wastewater treatment processes. *Microbial Pathogenis.* 2017a; 109: 305—312. DOI: 101,016/j.micpath.2017.06.001.

[4] Ibrahim C, Mehri I, Hammami S, Mejri S, Hassen A, Pierre P. Removal of human astroviruses from hospital wastewater by two biological treatment methods: natural oxidizing lagoons and rotating biodisks. *Desalination and Water treatment.* 2017b; 89: 287—296. DOI: 105,004/dwt.2017.21356.

[5] Jetten MSM, Horn SJ, Van Loosdrecht MCM. Towards a more sustainable municipal wastewater treatment system. *Water Science and Technology.* 1997; 35 (9): 171–180.

[6] Benyahya M, Bohatier J, Laveran H, Senaud J, Ettayebi M. Les virus des eaux usées et leur élimination au cours des traitements des effluents pollués. Wastewater viruses and their elimination during the treatment of polluted effluents [Wastewater viruses and their elimination during the treatment of polluted effluents. Wastewater viruses and their disposal during the treatment of polluted effluents]. *Biological year* 1998; 78: 95 – 105.

[7] Verbyla ME, Mihelcic JR. A review of virus removal in wastewater treatment pond systems. *Water Research* 2015; 71:107 - 124. DOI: 101, 016/j.watres.2014.12.031.

[8] Khan MB, Lee XY, Nisar H, Ng CA, Yeap KH, Malik AS. Digital image processing and analysis for activated sludge wastewater

treatment. *Advances Experimental Medicine and Biology* 2015; 823: 227-48. DOI: 101,007/978-3-319-10984-8_13.

[9] Sano D, Amarasiri M, Hata A, Watanabe T, Katayama H. Risk management of infectious viral diseases in wastewater reclamation and reuse: Review. *Environmental International.* 2016; 91: 220—229. DOI: 101,016/j.envint.2016.03.001.

[10] Hassen A, Mahrouk M, Ouzari H, Cherif M, Boudabous A, Damelincourt JJ. UV disinfection of treated wastewater in a large-scale pilot plant and inactivation of selected bacteria in a laboratory UV device. *Bioressources Technology.* 2000; 74: 141—150.

[11] Hassen A, Jedidi N, Kallali H, Ferchichi M, Ghrabi A, Chebbi F, Chefai A, Saidi N, Ennabli M. Elimination of indicators and heavy metals during domestic wastewater treatment in a semi-industrial pilot station. *Water Sciences Technologies* 1994; 27 (4): 34-41.

[12] US EPA. *Principles of Design and Operations of Wastewater Treatment Pond Systems for Plant Operators, Engineers, and Managers.* EPA/600/R-11/088, Cincinnati, Ohio, USA; 2011.

[13] US EPA. *Design Manual: Municipal Wastewater Stabilization Ponds.* EPA-625/1-83-015. Washington, DC, the USA; 1983.

[14] Jagger J. *Solar-UV Actions on Living Cells.* Praeger, New York, NY, USA; 1985.

[15] Hauduc H, Rieger L, Oehmen A, Van Loosdrecht MCM, Comeau Y, Heduit A, Vanrolleghem PA, Gillot S. Critical review of activated sludge modeling: state of process knowledge, modeling concepts, and limitations. *Biotechnology and Bioengineering.* 2013; 110 (1): 24–46. DOI: http://dx.doi.org/10.1002/bit.24624.

[16] Ni BJ, Yu HQ. Microbial products of activated sludge in biological wastewater treatment systems: a critical review. *Critical Reviews in Environmental Science and Technology.* 2012; 42: 187—223. DOI: https://doi.org/10.1080/10643389.2010.507696.

[17] Dai H, Chen W, Lu X. The application of multi-objective optimization method for activated sludge process: a review. *Water Sciences and Technology.* 2016; 73 (2): 223–235. DOI: 102,166/wst. 2015.489.

[18] Li J, Yang F, Li Y, Wong F, Chua HC. Impact of biological constituents and properties of activated sludge on membrane fouling in a novel submerged membrane bioreactor. *Desalination* 2008; 225: 356—365. DOI: 101,002/bit.24624.

[19] Mara DD. *Domestic Wastewater Treatment in Developing Countries*. Earth scan/James and James. London UK; 2003.

[20] Ghrabi A, Ferchichi M, Drakidès C. Treatment of wastewater by stabilization ponds—Application to Tunisian conditions. *Water Sciences and Technology*. 1993; 28 (10): 193–199.

[21] Ferchichi M, Ghrabi A, Grasmick A. Urban Wastewater treatment by trickling filters and rotating biological reactor. *Water Research*. 1994; 28 (2): 437–444. DOI: https://doi.org/10.1016/0043-1354 (94) 90,281-X.

[22] Turki Y, Mehri I, Lajnef R, Rejab AB, Khessairi A, Cherif H, Ouzari H, Hassen A. Biofilms in bioremediation and wastewater treatment: characterization of bacterial community structure and diversity during seasons in municipal wastewater treatment process. *International Journal of Environmental Science and Pollution Research*. 2017; 24 (4): 3519–3530. DOI: 101,007/s11356-016-8090-2.

[23] Daigger GT, Boltz JP. Trickling filters and trickling filter-suspended growth process design and operation: a state-of-the-art review. *Water Environment Research*. 2011; 83 (5): 388–404. DOI: 10.2175/106143010X12681059117211.

[24] Montuelle B. Application of aerobic degradation of organic matter in wastewater-microbiological aspects. In *Water Biology: Methods and Techniques*. Masson, Paris ; 1988.

[25] Salvado H, Gracia MP, Amigo JM. Capability of ciliated protozoa as indicators of effluent quality in activated sludge plants. *Water Research* 1995; 29: 1041—1050.

[26] Ben Said M, Abbassi MS, Gómez P, Ruiz-Ripa L, Sghaier S, Ibrahim C, Torres C, Hassen A. Staphylococcus aureus isolated from wastewater treatment plants in Tunisia: occurrence of human and

animal associated lineages. *J Water Health.* 2017; 15 (4): 638–643. DOI: 102,166/wh.2017.258.

[27] Kirby AE, Shi J, Montes J, Lichtenstein M, Moe CL. Disease course and viral shedding in experimental Norwalk virus and Snow Mountain virus infection. *Journal of Medical Virology* 2014; 86 (12): 2055–2064. DOI: 101,002/jmv.23905.

[28] Kirby AE, Teunis PF, Moe CL. Two human challenge studies confirm high infectivity of Norwalk virus. *Journal of Infectious Diseases* 2015; 211 (1): 166–167. DOI:10.1093/infdis/jiu585.

[29] Tjon GM, Coutinho RA, Van der Hoek A, Esman S, Wijkmans CJ, Hoebe CJ, Wolters B, Swaan C, Geskus RB, Dukers N, Bruisten SM. High and persistent excretion of hepatitis A virus in immune-competent patients. *Journal of Medical Virology.* 2006; 78: 1398-1405. DOI: 101,002/jmv.20711.

[30] Takahashi M, Tanaka T, Azuma M, Kusano E, Aikawa T, Shibayama T, Yazaki Y, Mizuo H, Inoue J, Okamoto H. Prolonged fecal shedding of hepatitis E virus (HEV) during sporadic acute hepatitis E: evaluation of infectivity of HEV in fecal specimens in a cell culture system. *Journal of Clinical Microbiology.* 2007; 45: 3671—3679. DOI: 101,128/JCM.01086-07.

[31] Nicand E, Teyssou R, Buisson Y. Le risque fécal viral en 1998 [Viral fecal risk in 1998]. *Virology*. 1998; 2: 103—116.

[32] Gray J, Iturriza-Gómara M. Rotaviruses. *Methods in Molecular Biology* 2011; 665: 325-55. DOI: 101,007/978-1-60761-817-1_18.

[33] Gibson K E. Viral pathogens in water: occurrence, public health impact, and available control strategies. *Current Opinion in Virology.* 2014; 4:50—57. DOI: 101, 016/j.coviro.2013.12.005.

[34] Butler AJ, Thomas MK, Pintar KD. Expert Elicitation as a Means to Attribute 28 Enteric Pathogens to Foodborne, waterborne, Animal Contact, and Person-to-Person Transmission Routes in Canada. *Foodborne and Pathogens and Disease*. 2015; 12 (4): 335–344. DOI: 101,089/fpd.2014.1856.

[35] Di Bartolo I, Pavoni E, Tofani S, Consoli M, Galuppini E, Losio MN, Ruggeri FM, Varisco G. Waterborne norovirus outbreaks

during a summer excursion in Northern Italy. *New Microbiology.* 2015; 38 (1): 109–112. PMID: 25,742,154.

[36] Liu L, Johnson HL, Cousens S, Perin J, Scott S, Lawn JE, Rudan I, Campbell H, Cibulskis R, Li M, Mathers C, Black RE. Child Health Epidemiology Reference Group of WHO and UNICEF. Global, regional and national causes of child mortality: an updated systematic analysis for 2010 with time trends since 2000. *Lancet.* 2012; 379: 2151—2161. DOI: 101,016/S0140-6736 (12)60560-1.

[37] Wazny K, Zipursky A, Black R, Curtis V, Duggan C, Guerrant R, Levine M, Petri WA, Jr Santosham M, Scharf R, Sherman PM, Simpson E, Young M, Bhutta ZA. Setting research priorities to reduce mortality and morbidity of childhood diarrheal disease in the next 15 years. *PLoS Medicine.* 2013; 10 (5): e1001446. DOI: 101, 371/journal.pmed.1,001,446.

[38] Bosch A, Pinto R M, Abad FX. Survival and transport of enteric viruses in the environment. In *Viruses in Food*, S M Goyal (Ed.). Food Microbiology and Food Safety Series. Springer, New York, NY; 2006. 151—187 p.

[39] Cauchi, Hyvrard, Nakache, Schwartzbrod, Zagury, Baron, Carre, Courtois, Denis, Dernat, Larbaigt, Derangere, Martigne, Seguret. Folder: the reuse of wastewater after purification. *Techniques, Science and Methods*, 1996 ; 2: 81-118.

[40] Baumont S, Camard J-P, Lefranc A, Franconie A. *Reuse of wastewater: health risks and feasibility in islandof France.* (*Réutilisation des eaux usées: Risques sanitaires et faisabilité en Île-de-France*). Report ORS (Rapport ORS), 2004 ; 220 p.

[41] Romero OC, Straub AP, Kohn T, Nguyen TH. Role of temperature and Suwannee River natural organic matter on inactivation kinetics of rotavirus and bacteriophages MS2 by solar irradiation. *Environmental Science and Technology* 2011; 45 (24): 10385e10393. DOI: 101,021/es202067f.

[42] Grabow WOK, Middendorff IG, Basson NC. Role of lime treatment in the removal of bacteria enteric viruses and coliphages in a

wastewater reclamation plant. *Appl. Environ. Microbiol.*, 1978; 35: 663-669. PMID: 25,621.

[43] Kott Y, Ben Ari H, Vinokur L. Coliphage survival as viral indicators in various wastewater quality effluents. *Progress in Water Technology*. 1978; 10: 337—346.

[44] Wallis C, Mehlnick JL. Cationic stabilization: a new property of enterovirus. *Virology*. 1962; 16: 504—506. PMID: 14,004,710.

[45] Wong K, Mukherjee B, Kahler AM, Zepp R, Molina M. Influence of inorganic ions on aggregation and adsorption behaviors of human Adenovirus. *Environmental Science and Technology*. 2012; 46 (20): 11145–11153. DOI: 101,021/es3028764.

[46] Templeton MR., Andrews RC, Hofmann R. Particle associated viruses in water: impacts on disinfection processes. *Critical Reviews in Environmental Science and Technology*. 2008; 38 (3): 137–164.

[47] Flint SJ, Enquist LW, Racaniello VR, Skalka AM. Principles of Virology, In: *Molecular Biology*. Third ed. volume 1. ASM Press: Washington, DC, the USA; 2009.

[48] Gutierrez L, Mylon SE, Nash B, Nguyen TH. Deposition and aggregation kinetics of rotavirus in divalent cation solutions. *Environmental Science and Technology*. 2010; 44 (12): 4552–4557. DOI: 101,021/es100120k.

[49] Girones R, Joffre JT, Bosch A. Isolation of marine bacteria with antiviral properties. *Canadian Journal of Microbiology*. 1989; 35: 1015—1021. PMID: 2,558,789.

[50] Ward RL, Knowlton DR., Winston PE. Mechanism of inactivation of enteric viruses in fresh water. *Applied and Environmental Microbiology*. 1986; 52: 450—459. PMID: 3,021,056.

[51] Iawprc. Study group on health-related water microbiology, bacteriophages as model viruses in water quality control. *Water Research*. 1991; 25: 529—545.

[52] Ben Yahya M, Laveran H, Bohatier J, Laveran H, Senaud J, Ettayebi M. Interactions between the ciliated protozoan Tetrahymena pyriformis and the simian Rotavirus-Sal 1. *European Journal of Protistology*. 1997; 33: 211—213.

[53] Danes L, Cerva L. The poliovirus and echovirus survival in Tetrahymena pyriformis culture in vivo. *Journal of Hygiene, Epidemiology Microbiology and Immunology* 1984; 28 (2): 193–200. PMID: 6088623.

[54] Scheid P, Schwarzenberger R. Acanthamoeba spp. As the vehicle and reservoirs of Adenoviruses. *Parasitology Research.* 2012; 111 (1): 479–485. DOI: 101,007/s00436-012-2828-7.

[55] Battistini R, Marcucci E, Verani M, Di Giuseppe G, Dini F, Carducci A. The ciliate-adenovirus interactions in experimental co-cultures of Euplotes octocarinatus and in wastewater environment. *European Journal of Protistology.* 2013; 49 (3), 381–388. DOI: 101,016/j.ejop. 2012.11.003.

[56] Kim T, Unno H. The role of microbes in the removal and inactivation of viruses in biological wastewater treatment systems. *Water Science and Technology.* 1996; 33: 243—250.

[57] Manage PM, Kawabata Z, Nakano S, Nishibe Y. Effect of heterotrophic nano-flagellates on the loss of virus-like particles in pond water. *Ecological Research.* 2002; 17: 473—479.

[58] Miki T, Jacquet S. Complex interactions in the microbial world: under-explored key links between viruses, bacteria and protozoan grazers in aquatic environments. *Aquatic Microbiol Ecology.* 2008; 51, 195–208.

[59] Cabbage CP, Gannon JJ, Cochran KW, Williams GW. Loss of infectivity of *Poliovirus* I in river water under simulated field conditions. *Water Research.* 1979; 13: 1091—1099.

[60] Hurst CJ. Fate of viruses during wastewater sludge treatment processes. *Critical Reviews in Environmental Control.* 1989; 18: 317—343.

[61] Kapuscinski RB, Mitchell R. Sunlight-induced mortality of viruses and Escherichia coli in coastal seawater. *Environmental Science and Technology.* 1983; 14: 1—6. DOI: 101,021/es00107a003.

[62] Lytle CD, Sagripanti J. Predicted inactivation of viruses of relevance to biodefense by solar radiation. *Journal of Virology.* 2005; 79 (22): 14244–14252. DOI: 101,128/JVI.79.22.14244-14252.2005.

[63] Bosshard F, Armand F, Hamelin R, Kohn T. Mechanisms of human adenovirus inactivation by sunlight and UVC light as examined by quantitative PCR and quantitative proteomics. *Applied and Environmental Microbiology.* 2013; 79 (4): 1325–1332. DOI: 101,128/AEM.03457-12.

[64] Kohn T, Nelson KL. Sunlight-mediated inactivation of MS2 coliphages via exogenous singlet oxygen produced by sensitizers in natural waters. *Environmental Science and Technology.* 2007; 41 (1): 192–197. PMID: 17,265,947.

[65] Nelson KL, Nguyen M, Schilling I. A novel pond design for more efficient disinfection. In: Oral Presentation, *10th IWA Specialist Conference on Wastewater Pond Technology.* International Water Association, Cartagena, Colombia. 2013.

[66] Mattle MJ, Vione D, Kohn T. A conceptual model and experimental framework to determine the contributions of direct and indirect photoreactions to the solar disinfection of MS2, phiX174, and adenovirus. *Environmental Science and Technology.* 2015; 49 (1): 334–342. DOI: 101,021/es504764u.

[67] Njiné T, Monkiédjé A, Nola M, Foko VS. Evaluation of bacterial and polluting loads of effluent from activated sludge wastewater treatment plants in Yaoundé, Cameroon. *Santé.* 2001 ; 11 (2): 79–84. PMID: 11,440,881.

[68] Sidhu JPS, Ahmed W, Palmer A, Smith K, Hodges L, Toze S. Optimization of sampling strategy to determine pathogen removal efficacy of activated sludge treatment plant. *International Journal of Environmental Science and Pollution Research.* 2017; 24 (23): 19,001–19,010. DOI: 101,007/s11356-017-9557-5.

[69] Campos CJA, Avant J, Lowther J, Till D, Leeds DN. Digital image processing and analysis for activated sludge wastewater treatment. Human norovirus in untreated sewage and effluents from primary, secondary and tertiary treatment processes. *Water Research.* 2016; 103: 224—232. DOI: 101,016/j. Wat. Res., 2016.07.045.

[70] Ali MA, El-Senousy WM, El-Hawaary SE. The enteroviruses in sewage: comparison of different technologies for wastewater

treatment and reuse. *Journal of the Egyptian Public Health Association.* 1997; 72 (5–6): 441–456. PMID: 17,214,145.

[71] Kitajima M, Iker BC, Pepper IL, Gerba CP Relative abundance and treatment reduction of viruses during wastewater. Treatment process identification of potential viral indicators. *Science of the Total Environment.* 2014; 488–489: 290–296. DOI: 101, 016/j.scitotenv. 2014.04.087.

[72] Prado T, Gaspar AM, Miagostovich MP. Detection of enteric viruses in activated sludge by feasible concentration methods. *Brazilian Journal of Microbiology.* 2014; 45 (1): 343–349.

[73] Qiu Y, Lee BE, Neumann N, Ashbolt N, Craik S, Maal-Bared R, Pang XL. Assessment of human virus removal during municipal wastewater treatment in Edmonton, Canada. *J Appl. Microbiol.* 2015; 119 (6): 1729–1739. PMID: 24,948,954.

[74] El-Senousy WM, Ragab AM, Handak EM. Prevalence of Rotaviruses Groups A and C in Egyptian Children and Aquatic Environment. *Food and Environmental Virology.* 2015; 7 (2): 132–141. DOI: 101,007/s12560-015-9184-6.

[75] Mondale B, Wraith MA. Use of shredded tire chips and tire crumbs as packing media in trickling filter systems for landfill leachate treatment. *Environmental Technology.* 2008; 29 (8): 827–836. DOI: 101,080/09593330801987566.

In: Wastewater Treatment
Editor: Adriana Magdalena
ISBN: 978-1-53616-370-4

Chapter 3

EXPLORING THE ENERGY POTENTIAL OF WASTEWATER WITH MICROBIAL FUEL CELLS

Brandon E. Oliphant, Stephen A. Caponetti, Pauline Sow, Jessica Boyer, Shivani Amin, Lauren Bahnsen and Birthe V. Kjellerup

Department of Civil and Environmental Engineering,
University of Maryland, College Park, MD, US

ABSTRACT

Wastewater treatment is an essential process that ensures the public health by promoting the health and safety of water systems. In the most general sense, wastewater treatment is the combination of processes like biological transformation, filtration, sedimentation, and disinfection, which are used to purify the wastewater. This chapter describes and discusses the most common wastewater treatment processes, the importance of the activated sludge process in wastewater treatment, and finally introduces the idea of implementing microbial fuel cells (MFCs) into the procedural design of wastewater treatment for resource recovery.

Microbial fuel cell technology is a wastewater treatment approach that uses microorganisms to convert chemical energy stored in the organic components of the wastewater into electrical energy. MFCs treat wastewater by taking advantage of respiring bacteria's ability to digest organics and produce free electrons. Current wastewater treatment approaches require significant energy input, however during these processes, the energy that is being transformed within the system is not captured but is instead released as waste energy (i.e., heat). This chapter reviews the feasibility of implementing microbial fuel cell technology into current wastewater treatment approaches and introduces the steps that need to be taken to utilize the electric energy released by bacteria and thus appropriately design the MFC process.

Depending on the type of wastewater being used, an MFC can be customized by choosing from a variety of anodes, cathodes, and cation specific membranes which can, in turn, promote energy production. MFCs have the potential to not only provide a sustainable wastewater treatment approach but they can also serve as alternative processes that are energetically beneficial to the user and specific to a variety of wastewater types, geographical locations, and environments. Therefore, this chapter concludes with a discussion on possible designs of industrial MFCs and comments on their respective benefits and drawbacks, in regard to industrial implementation.

1. Introduction

Most people with stable access to clean water usually do not concern themselves with where the water goes after they have finished with it. After use, the water, now considered "wastewater," is sent down a complex path of filtration, sedimentation, disinfection, and more. Domestic wastewater is composed of a plethora of different compounds that leave the household via the sewage system including fecal matter, food scraps, used water and household chemicals used for bathing or cleaning (Naidoo & Olaniran 2013). Each of these are composed of molecules from three main categories of organic compounds: lipids, carbohydrates and proteins, which all require different methods of treatment (Raunkjær, Hvitved-Jacobsen, & Nielsen 1994).

In some cases, wastewater can be reused without treatment, however depositing untreated wastewater into the environment can spread disease and cause damage to the environment and humans (Feenstra, Hussain, & van der Hoek, 2000). The effect that untreated wastewater can have on the environment is clear in places that have poor sewage management systems and that do not have wastewater treatment plants, such as rural locations and countries without regulated wastewater treatment. Some of the prominent, sub-lethal, effects that untreated wastewater can have on animals is endocrine disruption and oxidative stress response (Hashmi et al. 2018). Disruption of the endocrine system, generally caused by an influx of hormones, can have adverse effects on reproduction and can cause adverse physiological effects. Oxidative stress response, caused by reactive oxygen species, can cause cellular damage through the production of hydroxyl radicals and hydrogen peroxide. These components damage your cells by oxidizing important cellular components, such as nucleic acids or cellular proteins and membrane lipids (Camhi, Lee, & Choi 1995). Domestic wastewater also contains hormones, passed through human fecal matter, as well as reactive oxygen species, which can come from a number of the chemical components found in wastewater (Hashmi et al. 2018).

In order to minimize these health effects, most countries have rules and infrastructure in place for treating wastewater and controlling its release back into the environment. Water quality standards across the United States are governed by regulations, including the Clean Water Act, which regulate the quantity of contaminants that can be discharged into the environment (Grooms 2015). While the main aim of these regulations is to control the pollutant load deposited into the environment by industrial processes, they also regulate the content of the treated wastewater that is released back into the environment. One goal for the wastewater treatment industry is reaching the standards set by the legislation as efficiently as possible to keep the process affordable, while still releasing clean water.

This chapter will discuss the overall steps of a typical wastewater treatment processes, including biological transformation, filtration, sedimentation, aeration, and disinfection with emphasis on analysis of the activated sludge process and its applicability to the wastewater treatment process. As current wastewater treatment processes become outdated, it is important to implement new technologies as they become available. As a result, this chapter asserts that microbial fuel cells (MFCs) are a potential alternative or addition to current wastewater treatment technology. MFCs are energy efficient and environmentally sustainable. Moreover, with appropriate MFC designs and configurations, this technology can be used in conjunction with future generations of wastewater treatment technology.

2. Wastewater Treatment Methods

Current wastewater treatment methods comprise of a variety of operations that when combined, achieve high yields of treated water. There are three types of wastewater treatment: primary treatment, secondary treatment, and tertiary treatment. Primary treatment consists of the preliminary physical and chemical purification processes needed to separate the larger polluting particles from the water (Sonune & Ghate 2004). Secondary treatment often utilizes microorganisms to remove soluble contaminants such as nitrogen and carbon and insoluble pollutants such as phosphorus using chemical precipitation or alternatively biological phosphorus treatment (Gupta et al. 2012). Finally, tertiary treatment is applied to treat the water for safe human consumption using for instance ozonation or chlorination to eliminate fecal bacteria leftover from the previous treatment processes.

2.1. Primary Treatment

Primary treatment is used to remove settleable, suspended solids from the influent. One method is screening and filtration, which removes large quantities of waste from the influent, such as wood, hair, fecal solids, grease, and oils (Sonune & Ghate 2004). Another method, centrifugal separation, uses centrifugal devices to remove suspended solids from the water according to density properties (Gupta et al. 2012). Both of these methods are able to separate materials of larger size and greater density than water. The coagulation and flotation techniques are methods which can remove non-settleable solids. Coagulation involves the addition of a chemical that solidifies aqueous contaminants, whereas flotation removes floating pollutants via skimming (Gupta et al. 2012; Pescod 1992). Finally, the most prominent primary treatment method for wastewater treatment is the sedimentation and gravity separation process. In this process, the wastewater is left to settle, allowing suspended solids to gradually sink to the bottom of the tanks. This method is capable of converting one third of the biochemical oxygen demand (BOD), or measure of biodegradable organic matter remaining in the water, to solidified biosolids at the bottom of the tank (National Research Council 1996). These biosolids can further be used as fertilizer and the primarily treated water can be used for irrigation, enhancing the reusability of the water. Overall, primary treatment methods are capable of reducing 50-70% of the total suspended solids and 65% of the oils and greases of the wastewater (Pescod 1992). Although primary treatment is an effective clarifier of solid-containing water, secondary and tertiary treatments are needed to achieve potable water standards in most countries.

2.2. Secondary Treatment

Secondary treatment is used to remove soluble and insoluble pollutants that have progressed past the primary treatment stage through aerobic or anaerobic biological processes. When oxygen is present, the organic pollutants in the water go through aerobic decomposition via aerobic bacteria (Gupta et al. 2012). There are two prominent aerobic techniques, activated sludge and trickling filters. Trickling filters contain synthetic media of varying size that the wastewater is poured through (Pescod 1992). Bacteria congregates as biofilm on the media and consumes the organic matter of the wastewater. The activated sludge process involves a microorganism-wastewater suspension mix that is aerated to induce flocculation of the bacterial matter into a thick sludge (EPA 1998; National Research Council 1996). At this point the bacteria and organic matter (sludge) can be physically separated from the liquid water. As the process continues, some of the sludge is recycled into the next batch to be used as a nucleus for the bacteria and organic matter to cling to. As the amount of sludge grows, eventually the system reaches a state where it is producing more sludge than it needs for reuse. At this point the excess sludge is transferred to the biosolids treatment process stream for further treatment (Gillot et al. 2013).

Through these techniques, the concentration of biosolids in the water can be reduced by approximately 90% (Gupta et al. 2012). In contrast, without the use of aeration to introduce oxygen into the system, anaerobic bacteria convert the pollutants into simpler compounds which reduce the biological demand of the wastewater (Gupta et al. 2012). After these methods, the wastewater is typically disinfected and further treated with tertiary treatment processes to obtain water safe enough to be potable for humans.

2.3. Tertiary Treatment

Tertiary treatment is used to ensure that the discharged wastewater is of the quality required by the specific discharge permit that the wastewater treatment facility is operating under. The specifications of the permit vary based on the intended use for the water, be it for consumption or recycled use in another process. There are various approaches to tertiary treatment, some of the simplest and most common being evaporation, distillation, precipitation, and solvent extraction; however, more complex methods such as electrodialysis, ion exchange, and adsorption prove to be much more effective despite their increased complexity (Gupta et al. 2012).

Electrodialysis forces the passage of specific ions through a semipermeable membrane due to the presence of an electric current, thus allowing the solution to reach a greater level of purity (Gupta et al. 2012). In a similar manner, harmful ions in solution can be exchanged for harmless ions through the means of ion exchange with cation and anion exchangers (Gupta et al. 2012). As toxic ions are attracted to the active sites of these ion exchangers, the nontoxic ions of these exchangers replace the ones previously in solution, often resulting in a decrease of harmful organics and inorganics close to 95% (Gupta et al. 2012). It has also been studied that heavy metals and organic pollutants can be efficiently removed from solution through means of adsorption, a process in which atoms, ions, or molecules adhere to the surface of the absorbent which allows for the easy removal of pollutants from solution (Gupta et al. 2012). Without these valuable treatments processes completing the full cycle of wastewater treatment, very little water would be able to be recycled in usable and or potable water. However, the energy requirements of such wastewater treatment methods are quite taxing, forcing new changes to these processes to be explored in order to meet the world's growing energy demands.

3. Microbial Fuel Cell Technology

In the past decade, microbial fuel cells have gained increased visibility in the wastewater treatment sector, mainly because of their innovative designs and applications toward energy and cost savings as well as the production of clean energy specifically (Singh et al. 2010; Kocatürk-Schumacher et al. 2018). MFCs utilize the properties of microorganisms to function as catalysts, promoting the conversion of chemical energy into electrical energy through the oxidation and reduction of organic and inorganic matter in solution, typically consisting of sugars, organic acids, and biomass (Deval & Dikshit 2013). This matter is obtained from wastewater influent, making MFCs an ideal fit to wastewater treatment applications. The operating conditions of an MFC are simple, but their potential to function alongside modern wastewater treatment makes them very valuable.

Microbial fuel cells are electrochemical systems that use various types of bacteria to oxidize the organic matter in wastewater (Logan 2009). The bacteria metabolize the organics and produce free electrons, which are attracted to a lower potential environment. This generates an electrical current between the bacteria, located at the anode, and the lower potential environment at the cathode. The electric current flowing between the anode and cathode can be used as a source of power, allowing the chemical energy of the system to be harvested and utilized (Logan et al. 2006). It is important to note that the substrate solution (wastewater) must be replenished to allow the processes to take place continuously, a common practice in wastewater treatment plants (Gupta et al. 2012; Logan et al. 2006).

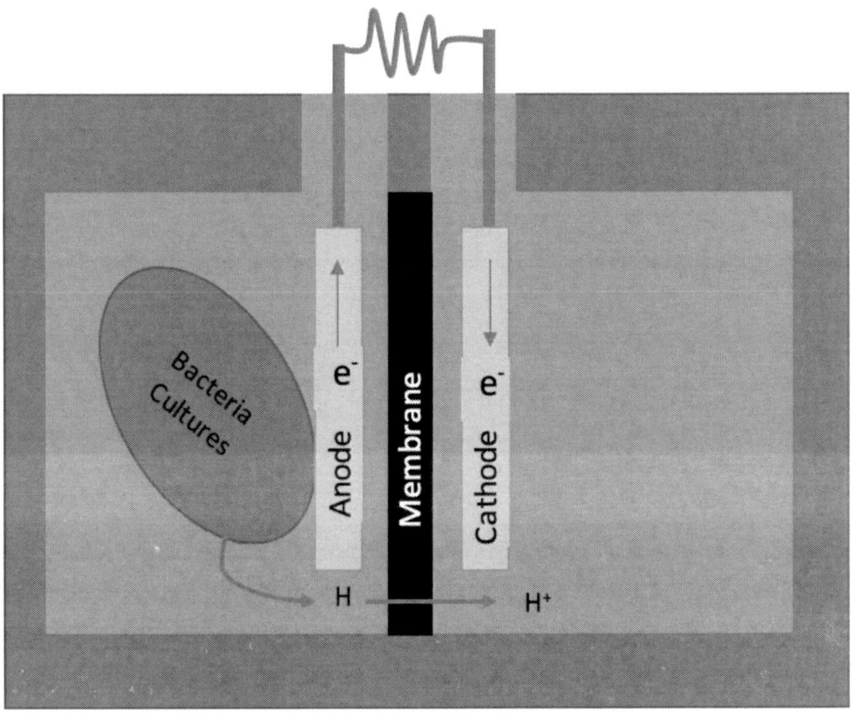

Figure 1. A model H-shaped MFC. Two chambers (anode and cathode) are separated by a PEM with wiring connecting the two electrodes. A resistor is positioned between the two electrodes in order to measure current and cell voltage as redox reactions occur within the cell (Freely after Logan et al. 2006).

At their most basic, MFC systems consist of two chambers: an anode chamber and a cathode chamber, along with a proton exchange membrane (PEM) or salt bridge (Rahimnejad et al. 2015). Salt bridges and PEMs serve the same purpose in a two-chambered MFC of providing a means of maintaining electrical neutrality within the system, but some studies have shown that there are benefits of implementing salt bridges in place of PEMs (Nair et al. 2013. Salt bridges are more cost effective than conventional PEMs due to their inexpensive and less complex designs (Nair et al. 2013; Deval & Dikshit 2013). Although a two-chambered MFC is not a universal design, this configuration prevents the unwanted mixing of anode and cathode solutions (combining solutions of electron acceptors and electron donors), which can result in an overall net loss of electrons (a loss of potential energy) (Deval & Dikshit 2013). The major component of an MFC's design is the selection of the PEM so that only protons are able to pass through the membrane, while substrates and electron acceptors/donors remain in their appropriate chambers (Logan et al. 2006; Pant et al. 2010; Liu & Logan 2004). This popular configuration of an MFC system is often referred to as an H-shaped system (Logan et al. 2006). Given their general ease of usability and simplistic design, these systems are particularly useful for experimenting with different materials, bacterial agents, and overall system sizes to optimize an MFC's electrical parameters such as cell voltage, potential, and power density (Park & Zeikus 2003).

Another foundational configuration of MFCs include single-chambered cells. These structures offer simpler designs and similar versatility of their construction. As mentioned before, a major area of study focuses on the importance and use of proton exchange membranes in conventional fuel cells. The presence of proton exchange membranes can affect the power density and overall electrical output of MFCs (Liu & Logan 2004). The use of an air cathode as well as the absence of a PEM has shown the capacity to promote the increase of power output from a system as well (Liu & Logan 2004). Platinum/Carbon cathodes or ones constructed from stainless steel or graphite brushes with titanium wire cores are also commonly applied. However, the presence of an air cathode promotes higher power densities in an MFC compared to more traditional architectural approaches with carbon-based electrodes and aqueous cathodes. (Call et al. 2009; Liu & Logan 2004). An MFC's cathode material does not significantly impact the system's overall power output, but a major factor in power density fluctuation is the surface area of the cathode; using metal catalysts has not shown enhancements in MFC energetic output (Parkash 2016; Call et al. 2009).

Within the anode chamber, microorganisms operate as active biocatalysts to produce an electrical current (Singh et al. 2010). These biocatalysts assume the role of oxidizing the organic matter to proliferate protons and electrons in solution. Electrons gather on the anode and flow through an electrical circuit to the cathode as protons are transferred through the PEM (Parkash 2016; Liu & Logan 2004). The protons then combine with electrons to form water at the cathode under aerobic conditions (Parkash 2016; Rahimnejad et al. 2015). Studies have shown that MFCs operating under conditions of mixed bacterial cultures produced larger power densities than the systems only utilizing pure bacterial cultures (Singh et al. 2010; Logan 2009; Logan et al. 2006). Several bacterial species have shown properties of electron transfer and recent research has shown that novel bacterial species display abilities to conduct interspecies electron transfer; some bacterial cells can transfer electrons directly between themselves instead of utilizing the anode surface thus supporting the theory of increased energy production/output in new MFC configurational designs (Kocatürk-Schumacher et al. 2018; Rahimnejad et al. 2015).

The configurations of MFCs are often evaluated under changing conditions such as temperature, pH, experiment time, wastewater volumes, electrode surface area and any other variables that may alter the efficiency and effectiveness of the MFC (Choudhury et al. 2017). The major areas of research for continued improvement of MFCs are electrode material, system configuration, and the anolyte and catholyte solutions. The anode is most often designed using conductive, chemically stable materials that are nonreactive with the electrochemically active bacteria and the compounds in the wastewater. A common versatile anodic material is carbon, which can be applied via graphite plates and rods. Carbon is also inexpensive and easy to handle thus making it an attractive material to use in MFCs (Choudhury et al. 2017; Logan et al. 2006). Graphite plates and rods have greater surface areas, which make them useful when optimizing energy output from the anode (the larger the surface area of the anode, the greater the energy output of the MFC) (Logan et al. 2006). For anode designs, ferricyanide has been applied in MFC studies due to its ability to operate near its open circuit potential while resisting reoxidation (Logan et al. 2006). Also, the ion exchange membrane used in MFCs can be varied to enhance the performance rates. Many MFC designs are constructed using an ion exchange membrane as a barrier between the anode and cathode chambers. Common membrane material includes Nafion and Ultrex CMI-7000 (Choudhury et al. 2017). However, a disadvantage to applying these membrane materials is that chemicals such as oxygen and organic matter can pass through the permeable material and enter the other chamber (Logan et al. 2006). As discussed here, although there are common materials used in the majority of MFC designs, the construction of an MFC is performed on a case-by-case basis; different circumstances warrant the use of different materials in order to ensure that the system is feasible, practical, affordable and thus useful for the advancement of MFC application for wastewater treatment.

A major component of developing MFC technology focuses on electrochemical specifications to reach enhanced parameters, such as power density and cell voltage as well as biological aspects such as the species of bacteria being used and how often the substrate solution is loaded into the system (Kocatürk-Schumacher et al. 2018; Rahimnejad et al. 2015). The overall performance of an MFC is determined by factors such as the presence of oxygen in the cathode chamber, the oxidation of organic substrates in the anode chamber, the transfer of electrons in solution, and the general permeability of the PEM. These factors can affect the overall usability and practicality of a given MFC design. Depending on the materials used and how they are orientated in an MFC, greater amounts of power can be generated, and current wastewater treatment processes can, in turn, be improved (Singh et al. 2010).

4. Implementing MFC Technology into Current Applications

There is currently a need for more environmentally friendly and sustainable methods for treating wastewater to combat the increase in fossil fuel use in the industry (Aelterman et al. 2006). MFCs present advantages over traditionally applied wastewater treatment approaches, since they have a low operating cost compared to current methods of wastewater treatment (Choudhury et al. 2017). However, MFCs have a high initial cost of establishment and still have some practical limitations restricting the commercial application (Santoro et al. 2017). These limitations need to be solved to facilitate MFC use towards energy generation with minimal energy input, cost requirements and environmental impact.

The majority of the research that has been conducted on wastewater treatment by MFC systems has been done at a laboratory-scale (Rozendal et al. 2008). This research was conducted to improve the understanding of the microbial communities on the anode and cathode, as well as develop improved cathode and anode materials, as well as catalysts (Hiegemann et al. 2018). The limited research that has been done on larger scale microbial desalination systems indicates that MFCs could be implemented into existing wastewater treatment facilities with great benefit (Rozendal et al. 2008). MFC technology could be used to reduce the overall amount of organic materials in the wastewater, as well as to reduce the nitrogen content of the wastewater. These two removal processes are the most costly for wastewater treatment plants (Leininger et al. 2019). Additionally, the electrical energy produced by MFCs could be used to undercut some of the energy requirements of the process, thus reducing the overall cost and environmental footprint.

Studies evaluating the integration of MFC technology into current wastewater treatment plants have shown that the performance of the MFC and the benefit to the wastewater treatment operation is dependent on where in the process stream MFCs are placed (Leininger et al. 2019). The current emphasis on MFC research focus on removing organic material (carbon and nitrogen) early in the wastewater treatment process. However, this might not be the most efficient use of the technology. MFCs could be applied to treat the wastewater at any stage from the primary effluent to the final stages of the process where liquids are produced as a byproduct from the biosolids handling processes (Leininger et al. 2019). Changing where the MFC is introduced into the process stream would influence the amount of organic material the MFC is treating which would in turn affect how much organic matter the system would process and in turn how much electricity the MFC could produce (Agricultural and Biosystems Engineering, Kassel University et al. 2017). The primary effluent has for example a lower concentration of organic material than the secondary effluent collected from the biosolids dewatering process, additionally the composition of the organic material could be different across different stages of the process stream.

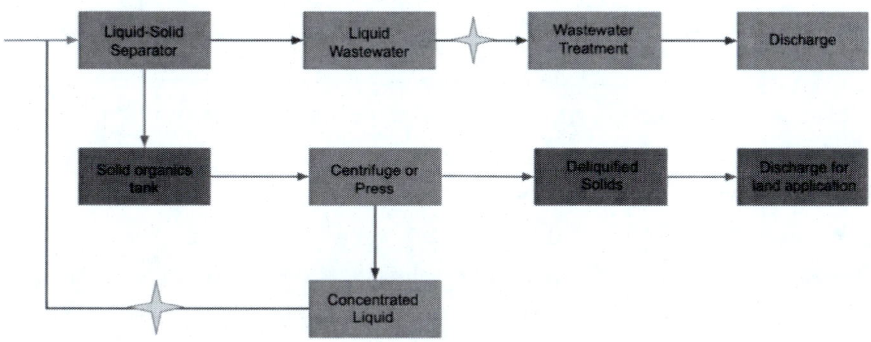

Figure 2. Example of a process stream from a wastewater treatment plant. The green stars indicate where an MFC were to be placed if it were to process primary effluent (top) or secondary effluent (bottom) (Freely after Leininger et al. 2019).

The higher concentration of organics in the secondary effluent makes it an attractive substrate for the MFC, however it also has a higher concentration of nitrogen in the form of nitrate and ammonium ions. The microbial communities that power the cell are sensitive to nitrogen concentrations and if the concentration of nitrate and ammonium are too high, they stop digesting the organic material and producing electrons (Leininger et al. 2019). One way to protect the bacteria from this is to dilute the high strength wastewater with some of the primary effluent. The resulting solution has higher organic content than the primary effluent alone, while maintaining nitrogen levels that are manageable for the bacteria to operate under.

Other factors that impact the MFC performance are the present microorganisms in the system, properties of the electrode materials, characteristics of the MFC design, and the process parameters. Further evaluation of these individual parameters and their collective influence may contribute to enhancement of the technology and acceleration of the commercialization (Choudhury et al. 2017). The main components of MFCs that affect overall cost and performance include the microorganisms and electrodes. Research in the field has found that microorganisms can adopt different metabolic processes resulting in different levels of energy output and power generation (Logan 2009; Kracke, Vassilev, & Krömer 2015). However, the electrode design is presently a challenge for designing cost effective MFCs (Choudhury et al. 2017). Modifications of the electrode material is an area that has shown improvements in the performance of MFCs, as improved materials can provide enhanced microbial attachment as well as electron transfer within the system (Choudhury et al. 2017). Further research in these areas will contribute to improvements upon MFCs, the implementation and future commercialization of MFC technology into wastewater treatment practices has the potential for growth and profit, resulting in a more cost effective and low energy system.

One of the challenges to developing large scale microbial fuel cell systems is the cost and performance of the cathode (Hiegemann et al. 2016), since the cathode is often the limiting factor for MFC performance. Each type of cathode has its own challenges. Biocathodes are challenging because cathodic biofilm is the least understood in terms of what conditions help it and hinder it (Hiegemann et al. 2016; Leininger et al. 2019). Laboratory scale tests have shown that catalysts such as platinum and various metal oxides are effective ways to lower the activation energy of the oxygen reduction reaction thus increasing the rate of the reaction and allowing the MFC to convert energy faster. At a smaller scale, catalysts are an effective way to speed the reaction, however the use of the catalysts at a large scale would be impractically expensive for an industrial operation.

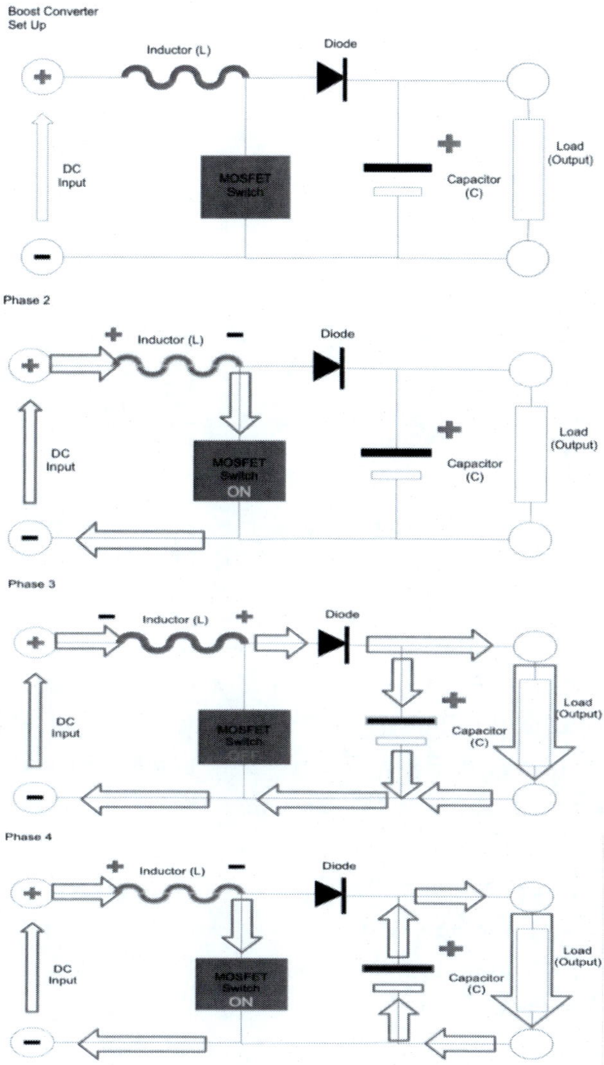

Figures 3–6. Green outline of a circuit component indicates charging, while a red outline indicates discharging. The color of the arrows indicates the source of the voltage, i.e., in Phase 2 the Inductor is being charged by the DC input and in Phase 3 the capacitor is being charged by both the discharging Inductor and the DC input (Freely from "Module 3.2 Boost Converters").

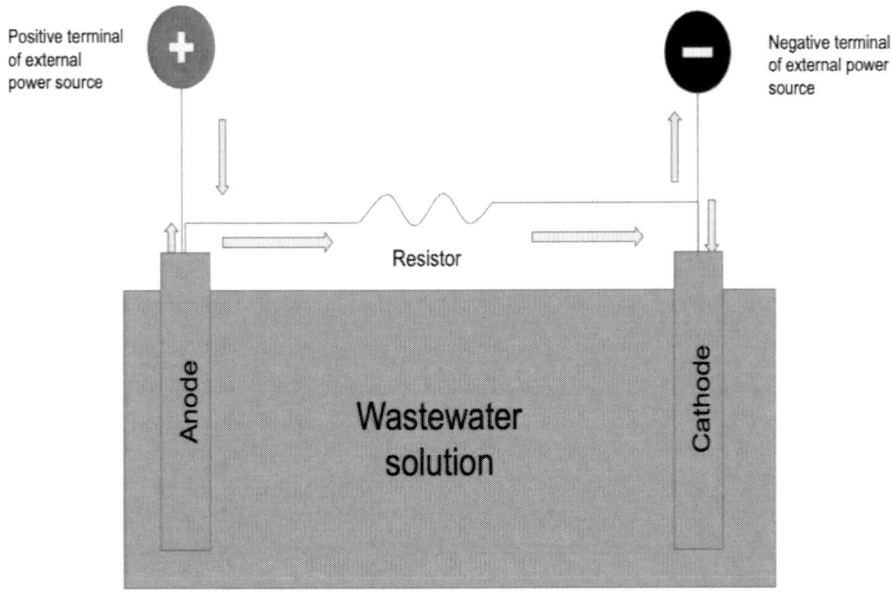

Figure 7. External power source creates areas of high potential at the anode and low potential at the cathode as electrons (yellow arrows) move around the circuit (Freely from Leininger et al. 2019).

While boost converters are a promising avenue for harvesting the voltage produced by MFCs, these converters have start up requirements for current and voltage that most MFCs cannot meet (Wang, Park, & Ren 2015). The solution is to use a charge pump between the MFCs and the boost converter to step up the MFC's voltage and current production to a level that meets the start-up requirement for the boost converter (Meehan, Hongwei, Gao, & Lewandowski 2009). Charge pumps are integrated circuits that can draw low current and use that current to charge a capacitor, which can then produce around 1.8V. This 1.8V can be boosted by the boost converter to a voltage that can be used to power different applications that draw voltage around 5V (Meehan, Hongwei, Gao, & Lewandowski 2009). A successful study conducted in 2015 showed this to be a practical way to harvest energy from MFCs. A system consisting of 96 2-L municipal wastewater MFC units (in eight rows of twelve) was evaluated. The system was utilizing between one and four of those rows in series and used a capacitor - boost converter power harvesting system to harvest up to 75 mA and 84-130 mW from the MFCs, which could run a small DC motor (Ge et al. 2015).

Another important aspect of the MFC that can affect the ability to treat wastewater is the microbial community that processes wastewater and produces the power. An effective biofilm for an MFC is composed of exoelectrogenic anode-respiring bacteria (ARB) such as G. *sulfurreducens*, which can metabolize simple organic molecules and produce electrons (Leininger et al. 2019). Bacteria can degrade small organics compounds such as acetate. However, wastewater contains many different and complex molecules that are not as easily transformed (Lee et al. 2008). A successful MFC biofilm contains bacteria that can perform hydrolysis to break down these molecules as well as bacteria that are capable of anodic respiration to produce electric energy in the form of electron movement (Leininger et al. 2019). To produce this microbial community, an external power source could be connected in parallel to the MFC circuit supplying a current, which induces electron movement between the anode and cathode. The electron movement initiated by the outside power source creates a potential difference between the anode (high potential) and the cathode (low potential).

The high potential anode serves as an electron sink, creating conditions in which it is thermodynamically favorable for the bacteria to release their electrons. Similarly, the lower potential cathode creates a thermodynamically favorable environment for the cathodic biofilm to accept the electrons. Creating such an environment encourages the bacteria to grow where they can degrade organic compounds and release electrons (anode) or where they can easily have access to electrons (cathode). Additionally, biofilms may be inoculated using an acclimation process in which the biofilm is initially formed using wastewater. Then the wastewater is replaced by a succession of acetate and glucose feedstocks, which serve to enrich the biofilm and enhance the bacterial activity, while they get acclimated to the organic rich wastewater. MFCs that were subjected to this acclimation process showed a higher coulombic efficiency than those which did not experience acclimation to the wastewater (Leininger et al. 2019).

CONCLUSION

Wastewater treatment is a critical process to ensuring the health of communities large and small. Traditional wastewater treatment protocol, comprised of primary, secondary and tertiary treatment steps, is a well-established means of converting wastewater into a usable product. As useful as processes like gravity filtration, activated sludge and reverse osmosis are, they are processes with high energy input requirements. Additionally, these processes offer few recoverable and useful byproducts, with the exception of reusable biosolids. With the addition of microbial fuel cell technology to wastewater treatment processes comes the opportunity to recapture energy from the system, as well as to minimize the required energy input. MFC technology operates on the chemical energy of the microbes already present in the wastewater. By starving the microbes of oxygen and providing a path of lower resistance than the wastewater environment that the microbes live in, the bacteria is encouraged to break down organic compounds in the wastewater, and release electrons through that path of least resistance. The movement of these electrons produces an electric current, completing the conversion of microbial chemical energy to usable electrical energy.

The efficiency of the MFC's electrochemical reaction is highly dependent on the organic content of the wastewater that it treats, which is in turn dependent on the stage of the wastewater treatment process stream that the water is harvested from. Additionally, MFC performance is a function of the configuration of the cell. Different cell setups like the H-shaped cell or the single-chambered cell, as well as variations in the cathode and anode materials all have different advantages and disadvantages that can affect the cells ability to treat the wastewater and recover energy. While energy recovery is one of the most valuable characteristics of the MFC, the technology does not lend itself to traditional energy recovery methods. Instead, a series of electrical components, like capacitors and inductors, must be used to effectively collect and redistribute the electrical energy of the system.

The MFC's ability to simultaneously treat wastewater and recover energy is an attribute that makes the technology invaluable as industrial processes place an increased emphasis on energy efficiency. As MFC technology continues to grow in scale, the likelihood of a feasible, practical, and worthwhile implementation of this technology into modern wastewater treatment processes grows as well. Despite the fact that the energy produced by MFCs has only been harnessed at the laboratory scale, the consistent variation of MFC design increases the capacity to improve and advance the technology. In time, microbial fuel cells may play a key role in wastewater treatment, potentially giving humanity another path to clean, renewable energy in tandem with recycled, reusable water.

References

Aelterman, P., K. Rabaey, P. Clauwaert, and W. Verstraete. 2006. "Microbial Fuel Cells for Wastewater Treatment." *Water Science and Technology* 54 (8): 9–15. https://doi.org/10.2166/wst.2006.702.

Agricultural and Biosystems Engineering, Kassel University, Mónica López Velarde Santos, Francisco Javier Rodríguez Valadéz, Centro de Investigación y Desarrollo Tecnológico en Electroquímica, Víctor Mora Solís, Centro de Investigación y Desarrollo Tecnológico en Electroquímica, Catalina González Nava, et al. 2017. "Performance of a microbial fuel cell operated with vinasses using different COD concentrations." *Revista Internacional de Contaminación Ambiental* 33 (3): 521–28. https://doi.org/10.20937/RICA.2017.33.03.14.

"*Bastre.Pdf.*" n.d. Accessed April 22, 2019. https://www3.epa.gov/npdes/pubs/bastre.pdf.

Call, Douglas F., Matthew D. Merrill, and Bruce E. Logan. 2009. "High Surface Area Stainless Steel Brushes as Cathodes in Microbial Electrolysis Cells." *Environmental Science & Technology* 43 (6): 2179–83. https://doi.org/10.1021/es803074x.

Camhi, S. L., P. Lee, and A. M. Choi. 1995. "The Oxidative Stress Response." *New Horizons (Baltimore, Md.)*3 (2): 170–82.

Choudhury, Payel, Uma Shankar Prasad Uday, Tarun Kanti Bandyopadhyay, Rup Narayan Ray, and Biswanath Bhunia. 2017. "Performance Improvement of Microbial Fuel Cell (MFC) Using Suitable Electrode and Bioengineered Organisms: A Review." *Bioengineered* 8 (5): 471–87. https://doi.org/10.1080/21655979.2016.1267883.

Council, National Research, Division on Earth and Life Studies, Commission on Geosciences Resources Environment and, and Committee on the Use of Treated Municipal Wastewater Effluents and Sludge in the Production of Crops for Human Consumption. n.d. *Use of Reclaimed Water and Sludge in Food Crop Production.* Accessed April 22, 2019. https://doi.org/10.17226/5175.

Deval, Animesh, and Anil Kumar Dikshit. 2013. "Construction, Working and Standardization of Microbial Fuel Cell." *APCBEE Procedia* 5: 59–63. https://doi.org/10.1016/j.apcbee.2013.05.011.

"*Developments in Wastewater Treatment Methods - ScienceDirect*." n.d. Accessed May 6, 2019. https://www.sciencedirect.com/science/article/pii/S0011916404003558?via%3Dihub.

"*Electricity Generation Using an Air-Cathode Single Chamber Microbial Fuel Cell in the Presence and Absence of a Proton Exchange Membrane. - PubMed - NCBI*." n.d. Accessed May 5, 2019. https://www.ncbi.nlm.nih.gov/pubmed/15298217.

Feenstra, Sabiena, Raheela Hussain, and Wim van der Hoek. 2000. "*Health Risks of Irrigation with Untreated Urban Wastewater in the Southern Punjab, Pakistan*." Institute of Publich Health, Lahore.

Ge, Zheng, Liao Wu, Fei Zhang, and Zhen He. 2015. "Energy Extraction from a Large-Scale Microbial Fuel Cell System Treating Municipal Wastewater." *Journal of Power Sources* 297 (November): 260–64. https://doi.org/10.1016/j.jpowsour.2015.07.105.

Gillot, Sylvie, Guenter Langergraber, Takayuki Ohtsuki, Andy Shaw, Imre Takacs, Stefan Winkler, and IWA Task Group on Modelling Practice. 2013. *Guidelines for Using Activated Sludge Models.* Edited by Leiv Rieger. Scientific and Technical Report 22. London [u.a]: IWA Publ.

Grooms, Katherine K. 2015. "Enforcing the Clean Water Act: The Effect of State-Level Corruption on Compliance." *Journal of Environmental Economics and Management* 73 (September): 50–78. https://doi.org/10.1016/j.jeem.2015.06.005.

Hashmi, Muhammad Arslan Kamal, Beate I. Escher, Martin Krauss, Ivana Teodorovic, and Werner Brack. 2018. "Effect-Directed Analysis (EDA) of Danube River Water Sample Receiving Untreated Municipal Wastewater from Novi Sad, Serbia." *Science of the Total Environment* 624 (May): 1072–81. https://doi.org/10.1016/j.scitotenv.2017.12.187.

Hiegemann, Heinz, Daniel Herzer, Edith Nettmann, Manfred Lübken, Patrick Schulte, Karl-Georg Schmelz, Sylvia Gredigk-Hoffmann, and Marc Wichern. 2016. "An Integrated 45 L Pilot Microbial Fuel Cell System at a Full-Scale Wastewater Treatment Plant." *Bioresource Technology* 218 (October): 115–22. https://doi.org/10.1016/j.biortech.2016.06.052.

Hiegemann, Heinz, Manfred Lübken, Patrick Schulte, Karl-Georg Schmelz, Sylvia Gredigk-Hoffmann, and Marc Wichern. 2018. "Inhibition of Microbial Fuel Cell Operation for Municipal Wastewater Treatment by Impact Loads of Free Ammonia in Bench- and 45 L-Scale." *Science of The Total Environment* 624 (May): 34–39. https://doi.org/10.1016/j.scitotenv.2017.12.072.

Kocatürk-Schumacher, Nazli Pelin, Joana Madjarov, Pavaris Viwatthanasittiphong, and Sven Kerzenmacher. 2018. "Toward an Energy Efficient Wastewater Treatment: Combining a Microbial Fuel Cell/Electrolysis Cell Anode with an Anaerobic Membrane Bioreactor." *Frontiers in Energy Research* 6 (September). https://doi.org/10.3389/fenrg.2018.00095.

Kracke, Frauke, Igor Vassilev, and Jens O. Kramer. 2015. "Microbial Electron Transport and Energy Conservation. "The Foundation for Optimizing Bioelectrochemical Systems." *Frontiers in Microbiology* 6 (June). https://doi.org/10.3389/fmicb.2015.00575.

Kumar Gupta, Vinod, Imran Ali, Tawfik A. Saleh, Arunima Nayak, and Shilpi Agarwal. 2012. "Chemical Treatment Technologies for Waste-Water Recycling—an Overview." *RSC Advances* 2 (16): 6380–88. https://doi.org/10.1039/C2RA20340E.

Lee, Hyung-Sool, Prathap Parameswaran, Andrew Kato-Marcus, César I. Torres, and Bruce E. Rittmann. 2008. "Evaluation of Energy-Conversion Efficiencies in Microbial Fuel Cells (MFCs) Utilizing Fermentable and Non-Fermentable Substrates." *Water Research* 42 (6–7): 1501–10. https://doi.org/10.1016/j.watres.2007.10.036.

Leininger, Aaron, Wing-Mei Ko, Mark Ramirez, and Birthe V. Kjellerup. 2019. "Implementation of Upscaled Microbial Fuel Cells for Optimized Net Energy Benefit in Wastewater Treatment Systems." *Journal of Environmental Engineering* 145 (5): 04019020. https://doi.org/10.1061/(ASCE)EE.1943-7870.0001531.

Liu, Hong, and Bruce E. Logan. 2004. "Electricity Generation Using an Air-Cathode Single Chamber Microbial Fuel Cell in the Presence and Absence of a Proton Exchange Membrane." *Environmental Science & Technology* 38 (14): 4040–46. https://doi.org/10.1021/es0499344.

Logan, Bruce E. 2009. "Exoelectrogenic Bacteria That Power Microbial Fuel Cells." *Nature Reviews Microbiology* 7 (5): 375–81. https://doi.org/10.1038/nrmicro2113.

Logan, Bruce E., Bert Hamelers, René Rozendal, Uwe Schröder, Jürg Keller, Stefano Freguia, Peter Aelterman, Willy Verstraete, and Korneel Rabaey. 2006. "Microbial Fuel Cells: Methodology and Technology [†]." *Environmental Science & Technology* 40 (17): 5181–92. https://doi.org/10.1021/es0605016.

Meehan, A., Hongwei Gao, and Z. Lewandowski. 2009. "Energy Harvest with Microbial Fuel Cell and Power Management System." In *2009 IEEE Energy Conversion Congress and Exposition*, 3558–63. San Jose, CA: IEEE. https://doi.org/10.1109/ECCE.2009.5316034.

"*Module 3.2 Boost Converters*." n.d. Eric Coates MA BSc. http://www.learnabout-electronics.org/PSU/psu32.php.

Naidoo, Shalinee, and Ademola Olaniran. 2013. "Treated Wastewater Effluent as a Source of Microbial Pollution of Surface Water Resources." *International Journal of Environmental Research and Public Health* 11 (1): 249–70. https://doi.org/10.3390/ijerph110100249.

Nair, Ramya, K. Renganathan, Barathi S., & K. Venkatraman. 2013. "Performance of Salt-Bridge Microbial Fuel Cell at Various Agarose Concentrations Using Hostel Sewage Waste as Substrate." *International Journal of Advancements in Research & Technology* 2 (5) ISSN 2278-7763. Accessed April 22, 2019. http://www.ijoart.org/papers/ Performance- of- salt- bridge- microbial- fuel- cell- at- various-agarose-Concentrations-using-hostel-sewage-waste-as-substrate.html.

Pant, Deepak, Gilbert Van Bogaert, Ludo Diels, and Karolien Vanbroekhoven. 2010. "A Review of the Substrates Used in Microbial Fuel Cells (MFCs) for Sustainable Energy Production." *Bioresource Technology* 101 (6): 1533–43. https://doi.org/10.1016/j.biortech.2009.10.017.

Park, Doo Hyun, and J. Gregory Zeikus. 2003. "Improved Fuel Cell and Electrode Designs for Producing Electricity from Microbial Degradation." *Biotechnology and Bioengineering* 81 (3): 348–55. https://doi.org/10.1002/bit.10501.

Parkash, Anand. 2016. "Microbial Fuel Cells: A Source of Bioenergy." *Journal of Microbial & Biochemical Technology* 8 (3). https://doi.org/10.4172/1948-5948.1000293.

Pescod, M.B. n.d. "Wastewater Treatment and Use in Agriculture." *FAO*. Accessed April 22, 2019. http://www.fao.org/3/t0551e/t0551e05.htm#3.2.2 primary treatment.

Rahimnejad, Mostafa, Arash Adhami, Soheil Darvari, Alireza Zirepour, and Sang-Eun Oh. 2015. "Microbial Fuel Cell as New Technology for Bioelectricity Generation: A Review." *Alexandria Engineering Journa l*54 (3): 745–56. https://doi.org/10.1016/j.aej.2015.03.031.

Raunkjær, Kamma, Thorkild Hvitved-Jacobsen, and Per Halkjær Nielsen. 1994. "Measurement of Pools of Protein, Carbohydrate and Lipid in Domestic Wastewater." *Water Research* 28 (2): 251–62. https://doi.org/10.1016/0043-1354(94)90261-5.

Rozendal, René A., Hubertus V.M. Hamelers, Korneel Rabaey, Jurg Keller, and Cees J.N. Buisman. 2008. "Towards Practical Implementation of Bioelectrochemical Wastewater Treatment." *Trends in Biotechnology* 26 (8): 450–59. https://doi.org/10.1016/j.tibtech.2008.04.008.

Santoro, Carlo, Catia Arbizzani, Benjamin Erable, and Ioannis Ieropoulos. 2017. "Microbial Fuel Cells: From Fundamentals to Applications. A Review." *Journal of Power Sources* 356 (July): 225–44. https://doi.org/10.1016/j.jpowsour.2017.03.109.

Singh, D., D. Pratap, Y. Baranwal, B. Kumar, and R. K. Chaudhary. 2010. "Microbial Fuel Cells: A Green Technology for Power Generation." *Annals of Biological Research* 1 (3): 128–38.

Sonune, Amit, and Rupali Ghate. 2004. "Developments in Wastewater Treatment Methods." *Desalination, Desalination Strategies in South Mediterranean Countries*, 167 (August): 55–63. https://doi.org/10.1016/j.desal.2004.06.113.

Sun, Jian, Wenjing Xu, Yong Yuan, Xingwen Lu, Birthe V. Kjellerup, Zhenbo Xu, Hongguo Zhang, and Yaping Zhang. 2019. "Bioelectrical Power Generation Coupled with High-Strength Nitrogen Removal Using a Photo-Bioelectrochemical Fuel Cell under Oxytetracycline Stress." *Electrochimica Acta* 299 (March): 500–8. https://doi.org/10.1016/j.electacta.2019.01.036.

Wang, Heming, Jae-Do Park, and Zhiyong Jason Ren. 2015. "Practical Energy Harvesting for Microbial Fuel Cells: A Review." *Environmental Science & Technology* 49 (6): 3267–77. https://doi.org/10.1021/es5047765.

Wang, Liang, Jinli Liu, Quanyu Zhao, Wei Wei, and Yuhan Sun. 2016. "Comparative Study of Wastewater Treatment and Nutrient Recycle via Activated Sludge, Microalgae and Combination Systems."

Bioresource Technology 211 (July): 1–5. https://doi.org/10.1016/j.biortech.2016.03.048.

Xin, Xiaodong, Junming Hong, Junguo He, and Wei Qiu. 2019. "An Integrated Approach for Waste Activated Sludge Management towards Electric Energy Production/Resource Reuse." *Bioresource Technology* 274 (February): 225–31. https://doi.org/10.1016/j.biortech.2018.11.092.

In: Wastewater Treatment
Editor: Adriana Magdalena
ISBN: 978-1-53616-370-4

Chapter 4

TREATMENT AND USES OF BIOGAS AND BIOETHANOL WASTEWATER

***Fábio Spitza Stefanski*[1], *Thamarys Scapini*[1], *Aline Frumi Camargo*[1], *Caroline Dalastra*[1], *Natalia Klanovicz*[1], *Karina Paula Preczeski*[1], *Fabiane Czapela*[1], *Simone Kubeneck*[1], *Gislaine Fongaro*[2] and *Helen Treichel*[1,*]**

[1]Laboratory of Microbiology and Bioprocesses, Federal University of Fronteira Sul, Erechim, RS, Brazil
[2]Laboratory of Virology, Department of Microbiology, Immunology and Parasitology, Federal University of Santa Catarina, Florianópolis, SC, Brazil

ABSTRACT

Renewable fuels, such as biogas and bioethanol, represent main strategies in the energy sector to replace dependence on fossil fuels. However, both processes link the production of biofuels, as well as

* Corresponding Author's E-mail: helentreichel@gmail.com.

combine with the formation of wastewater inherent to production stages that affect the concepts of green economy and pose risks to the environment. In this way, post-treatment processes prior to release in nature are required to ensure the depollution of the wastewater. In this sense, this chapter seeks to cover strategies aimed at minimizing the expense of water during every stage of energy and biofuel production, as well as forms of reuse and recycling associated with treatments that guarantee the utilization of wastewater in order to develop the circular economy in biogas and bioethanol plants finding ways for a more eco-friendly energy production.

Keywords: biofuels, circular economy, clean energy

Biogas and Bioethanol: Alternative Fuels

Over the years, water use has been increasingly demanded human activities that are constantly growing, such as agriculture, industrialization in several sectors, and energy consumption, especially with regard to the use of biofuels that demand high amounts of water to obtain those (Gheewala et al. 2013).

Currently, biofuels have gained more and more space in the energy sector due to the damage caused to the environment by fossil fuels, the main current energy sources (Cardona and Sánchez 2007). There is a considerable increase in the demand for renewable sources in the coming years, mainly in relation to the production of bioethanol (Chavez-Rodrigues and Nebra 2010) and biogas (Chen, Chen and Song 2012).

In the case of biogas, the various applications of anaerobic digestion go far beyond just generating electricity. It is used as vehicular fuel, electricity and heat cogeneration, treatment and injection in gas networks, for domestic heating, chemical industry, among others (Al Seadi et al. 2008). Studies such as de Chen et al. (2016) and Sibilo et al. (2017) evaluated the benefits of using biogas in the practice of trigeration (combined cooling, heating and power generation) in different parts of the world.

Biogas use has been gaining strength due to the vast possibility of the use of residues that can serve as raw material for this type of technology. According to Świątczak, Cydzik-Kwiatkowska and Zielińska (2019) fermentation fit as an important technology to achieve the objective of the European Commission aiming to produce 20% of all energy consumed using renewable resources by 2020. Anaerobic digestion comprises a biological process in which microorganisms decompose organic matter into methane and carbon dioxide under anaerobic conditions (without oxygen) (Chandler et al. 1980). Among the organic matter types, biomass from animal manure exhibits an energetic potential projected to range from 1.2 x 10^3 to 2.3 x 10^3 PJ/year in 2030 in the European Union countries 28 (Meyer et al. 2017). Agricultural residues, vinasse, effluent treatment sludge, and urban solid waste also present potential as raw material for the anaerobic digestion (Moreda 2016).

Biogas chain is an interesting approach to reduce the pressures of energy shortage and emissions of greenhouse gases. However, it needs attention to the consumption of water resources during the digestion process (Zhao, Cheng and Yang 2014). Recent studies such as de Bansal, Tumwesige and Smith (2016), consider the potential of domestic water recycling, rainwater harvesting, and aquaculture to provide needed water during the digestion of biogas in different countries of sub-Saharan Africa region that is classified as problematic in relation to water scarcity. The products resulting from the fermentation can still contribute to the environmental aggravation if they do not receive due attention. Typically, the formed digestate requires post-treatments (physicochemical, the most used) that guarantee greater safety in the disposal or subsequent use of this material (Törnwall et al. 2017). However, the digested can be divided into a liquid and solid phase. The solid phase is generally intended for use as fertilizers but is considered putative because of the risk of eutrophication arising from the migration of particulate nutrients (Nkoa 2014). The liquid phase, however, requires more attention than the solid phase because of the abundant concentration of soluble ammonium, which hinders the process of management of this material (Xia and Murphy 2016). The use of specific treatments can produce high-quality effluents that can be reused for irrigation or for recreational purposes (Bixio et al. 2006).

The constant search for the increase of the efficiency of this type of technology and the management and treatment of generated effluents reached sidereal proportions. Dhoble and Pullammanappallil (2014) studied the effect of several factors on a two-stage reactor design that separated the soluble components from the solids and treated them separately. According to them, the wastes generated during human spaceflight are materials of no further use and require storage until the end of missions, making it virtually impossible to obtain all wastes stored during long-term space travel. Because of this, the authors proposed a prototype digester sized for four crewmembers for long-term exploratory missions. The system not only increased degradation but also produced 20-50% more methane than the mixed (liquid and solid) system, which could still be useful in providing fuel during the exploration (Dhoble and Pullammanappallil 2014).

In the case of bioethanol, it is known by different generations that vary according to the energy source used (Mohapatra, Ray and Ramachandran 2019). Its production is an essential part of the sector, which, in addition to first-generation biofuels (grains, sugar beet, and sugar cane), is investing in studies and expansion in the world to produce second-generation biofuels (cellulosic ethanol) and third parties (ethanol from algae) (Araújo 2016).

Each generation has a different process of producing ethanol, as well as the substrate used. First-generation (1G) ethanol is produced from the direct conversion of fermentable sugars, such as glucose, sucrose, and fructose, into ethanol during a microbiological fermentation process. The most commonly used raw materials for 1G ethanol are starch-based, such as corn, wheat and cassava, which are sources rich in polysaccharides that are hydrolyzed to provide fermentable sugars, or sugar-based raw materials such as sugarcane, sugar, beet and sorghum, which are rich in fermentable sugars (Vohra et al. 2014; Azhar et al. 2017; Bechara et al. 2018). The 1G ethanol production system from sugar-based raw materials is based on washing processes to remove impurities from the crop. This material goes through the grinding and diffusion process to extract the fermentation broth, which will carry the sugars necessary for conversion into ethanol, and finally, this broth will continue to the fermentation process, where will be added the microorganisms responsible for the conversion of sugars into ethanol (Bechara et al. 2018). The yeast commonly used in large-scale processes is *Saccharomyces cerevisiae*, by the wide adaptation to industrial conditions, such as tolerance to high temperatures, broad pH range, osmotic stress and high levels of ethanol (Balakumar and Arasaratnam 2012). After the fermentation process, purification processes with distillation and dehydration are performed (Mosqueira-Salazar et al. 2013). In the 1G process by amylaceous raw material saccharification processes are used, for conversion of polysaccharides into fermentable sugars (Bechara et al. 2018).

Second-generation (2G) ethanol production processes raw materials from non-food sources or lignocellulosic biomass are used. The main difference between the 1G and 2G systems is the technological route. In the 2G ethanol producing is need the pre-treatment processes before the fermentation, because of the complex structure of raw material. This process disintegrates the structure of the materials and thus increased accessibility to sugars for conversion into ethanol (Liu et al. 2019). It should be noted that the step preceding the fermentation processes to obtain fermentable sugars is the main difference between the ethanol production processes from simple sugar, starch or lignocellulosic material (Vohra et al. 2014, Azhar et al. 2017).

For the production of third generation bioethanol (3G), the use of microalgae and macroalgae is used as source of raw material. They can undergo two types of hydrolysis, acidic and enzymatic, where acid hydrolysis aims to hydrolyze algae in sugars with the use of sulfuric acid (Jang et al. 2012) and the enzymatic hydrolysis that uses cellulases to degrade polysaccharides (Jambo et al. 2016) obtaining sugars for fermentation. The fermentation process for this generation can be done in two ways. One is called the conventional method in which the sugars are obtained in the hydrolysis to be used in the fermentation (Jambo et al., 2016). The other one consists in the hydrolysis and the fermentation in a single step, where the substrate used, the enzyme and the yeast are arranged together so that the sugars obtained in the hydrolysis are already converted into bioethanol (Jambo et al., 2016). In the stages before the production of ethanol in industry, first-generation ethanol has the largest water footprint and the highest carbon production due to irrigation and planting stages (Yang and Chen 2013; Cardozo, Bordonal and La Scala Jr 2016; Oker et al. 2018). Regarding 2G ethanol, there is a lower water footprint than the previous generation, due to the use of lignocellulosic residues from crops destined for the food or energy sector, not directly using the main portion of the crop (Carrillo-Nieves et al. 2019; Tyagi et al. 2019, 22-24; Mathioudakis et al. 2017). Coming to 3G and 4G ethanol, we find an even lower water footprint, since the raw material for this generation is not intended for food, but is geared towards biofuels. For algae production, it is possible to use wastewater from industrial processes, jointly processing liquid waste and producing bioethanol feedstock (Gouveia 2011; Dutta, Daverey and Lin 2014).

During 2G, 3G and 4G ethanol is observed less water footprint than 1G ethanol. For example, to produce corn ethanol are spent 10-324 gallon of water per gallon of ethanol produced while for switchgrass bioethanol production are spent 1.9-9.8 gallon of water per gallon of bioethanol produced. This calculation includes all the water used in the conversion to the ethanol product, including irrigation (Wu and Chiu 2011). Even so, in all generations large volumes of water are required for the steps of imbibition, hydrolysis, fermentation, and distillation, generating a large amount of effluent that could be treated and returned to the process. In addition to the water expenditure in the steps mentioned above, water is still required for the cooling and condensation processes inside the industrial plant, water that can also be derived from reuse water (Pina et al. 2017; Siriwong et al. 2019). After obtention of the raw material, in the industrial stage, several processing steps are required, which requires the expenditure of water to obtain bioethanol, which may be potable water or wastewater (Pina et al. 2017). It is important to invest in the industries in alternatives that aim at the reuse of resources and prioritize the closed cycle, so that the volume of water used in the process is decreasing, increasing water security and decreasing the water footprint.

Considering these aspects, new technologies for bioenergies will be a high potential in the future, however, they lack strategies to promote the adequate management of water consumption throughout the production cycle, since there is great concern about the surplus wastewater generated of the industrial stage and that can affect the environment. In this sense, this chapter show aspects related to the process of the biogas and bioethanol industries with an interest in the reduction of the consumption of noble water seeking strategies with the use of adapted microorganisms or integrated technologies that aim at the reduction of water uses, promoting recycle and reuse water from the process and energy chain.

WATER USES IN BIOGAS AND BIOETHANOL PLANTS, WASTEWATER GENERATED AND REDUCTION STRATEGIES

Biogas

Anaerobic digestion follows transformation processes from organic biomass that through the microbiological action results in biogas and effluent, the two nutrient-rich parts (Lantz et al. 2007; Srinivasan 2008; Freitas et al. 2019). At the end of processing the anaerobic digestion two products are considered to be of high added value, the biogas that is composed mainly by methane (CH_4) and carbon dioxide (CO_2) (among other compounds) and the digestate, a sludge-like mixture, which is divided into solid and liquid portions (Ngumah et al. 2013; Felca et al. 2018).

The composition of the substrates used in the anaerobic digestion process is important for the biogas yield due to the number of nutrients available to the microorganisms, besides stabilizing the process (Angelidaki and Ellegaard 2003).

The use of agricultural substrates for the production of biogas has generated a competition of these, since the use for this purpose ends up diverting them from their initial use, such as for animal feed production

(Van Stappen et al. 2016). Organic waste and materials with high levels of lignin, cellulose, and hemicellulose can also be used in the anaerobic digestion process, but they need to be pre-treated in order to increase the digestion capacity (Al Seadi et al. 2008; Đurđica et al. 2019).

Thus, the use of animal manure as a substrate is a good alternative for the production of bioenergy since it does not compete with other purposes, does not need previous treatment, contributes to the reduction of greenhouse gases emissions and still presents a wide variety of nutrients necessary for microbial growth (Angelidaki and Ellegaard 2003; Esteves et al. 2019). However, animal manure as a single substrate in the anaerobic digestion process has low methane production efficiency due to high water and fiber content (Angelidaki and Ellegaard 2003). The anaerobic co-digestion of two or more substrates is more efficient when compared to the addition of only one type of substrate to the biodigester, because it increases the yield of the biogas, since the mixture presents a nutritional balance adequate to the development of the microorganisms (Angelidaki and Ellegaard 2003).

Khayum, Anbarasu, and Murugan (2018) studied biogas production through the co-digestion of spent tea waste (STW) with cow manure (CM). The substrates were mixed in five different proportions and kept in different anaerobic digesters, on a laboratory scale. The raw materials had a moisture content of 71.2% and 9.36% for CM and STW, respectively. The best biogas production condition was observed in the co-digestion of the substrates containing 30% STW and 70% CM during 25 days of digestion with a yield of 0.82 mL/g VS (volatile solids). The authors also observed that mixing the two substrates attains a significant increase (127%) in the production of biogas compared to the digestion of a single substrate (CM). Bouallagui et al. (2009) studied the performance of biogas production by co-digestion of fruit and vegetable residues (FWV) with fish waste (FW), abattoir wastewater (AW) and waste activated sludge (WAS). A percentage of the residues were mixed in one volume of water and mixed with the FVW. The experiments were conducted in four different laboratory-scale anaerobic reactors. The addition of AW and WAS to FVW increased biogas production by 51.5% and 43.8%, respectively, demonstrating that the addition of these substrates can significantly increase biogas yield when compared to single FVW digestion.

Productive chain of biogas is attractive due to several advantages that the process presents, among them the reduced amount of water that is necessary during anaerobic digestion. This aspect is directly related to versatility that process presents for the use of different raw materials, such as wet biomass (moisture content higher than 60%), for example, animal manure, sludge, and processing sludge, as well as drier substrates, as agricultural residues (especially grains) (Al Seadi et al. 2008).

The technology used in biodigesters depends on geographic, climatic factors and energy availability. In this context, small-scale biodigesters are generally cheap, easy to operate and robust, the Chinese and Indian models represent the category well. The biodigesters that incorporate the integrated system treatment of organic residues and recycling of nutrients bring a concept of a circular economy, providing environmental and economic benefits for the energy and agricultural sector (Al Seadi et al. 2008). On the way, upflow reactors (UASB) are widely used for the treatment of organic waste of animal origin and biogas production. This type of biodigestor allows the recirculation of the effluent through its upward flow and the insoluble part of the effluent tends to be immobilized in the sludge blanket. This process causes the reduction of the concentration of organic matter to very low levels, as well as the time of retention, which is attractive in the cost-effective sense of the digester (Álvarez et al. 2004; Silva Jr et al. 2016; Freitas et al. 2019).

In the current scenario, strategies that follow the circular economy model receive great promotions from the approximation of the environmental and economic areas. The aim of a circular economy is to reduce the entry of virgin natural resources and the generation of waste, thus closing a cycle in which input resources are equivalent to output, following a sustainable flow (Islam 2017; Manninen et al. 2018; Ferella et al. 2019). Life cycle studies of these processes aim to find improvement approaches, in the case of biogas there are technologies related to the reduction of carbon footprint and energy. Few studies deal with the analysis of water use in this specific process, the focus is on feedstock, since it employed great influence in the course of the process (Leonzio 2016; Castellani et al. 2017; Ferella et al. 2019).

The implementation of the circular economy in anaerobic digestion reduces costs, especially those referring to the choice of substrate since they are chosen to aim at the reuse or recycling of waste. In these cases, the cogeneration also presents continuation in the scope of the circular economy (Ferella et al. 2019). The perspective of the circular economy presents a series of advantages; especially for the cattle raising activity that generates a large number of residues, therefore, the appropriate treatment of these residues avoids contamination of water bodies, groundwater, bringing environmental and health security to the population (Freitas et al. 2019).

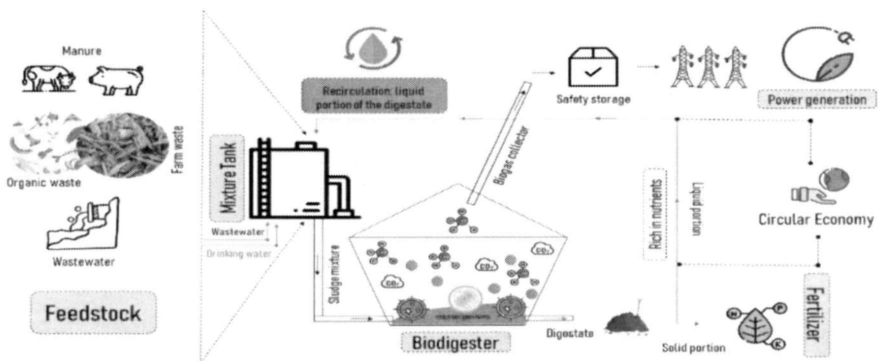

Figure 1. Flow the inputs and outputs of resources along the chain of a biogas plant.

Following the perception of a circular economy, a strong trend is the reuse of all possible resources, which includes wastewater recycling (Puyol et al. 2017; Lu et al. 2019). The wastewater from the anaerobic digestion process has a high organic nutrient load, considering treatment for this residual the economic and energy cost would be significant (Kulakov and Lebedeva 2011; Gogina and Makisha 2014; Scherbakow et al. 2015; Makisha 2016). Thus, the reuse of wastewater in a process that generated it aggregates in such a way that gains in terms of ecological and economic efficiency, low generation of effluents and emissions and finally, energetic independence, that is, the process is self-sufficient (Figure 1) (Makisha 2016).

Studies show a cubic meter of wastewater has the potential to generate about 3 to 6 megajoules of thermal energy, according to the concentration of organic waste in the effluent. Taking into account that for the maintenance of the biogas plant about 2.4 megajoules are used, which represents that energy generated by wastewater presents greater potential when reused and that a biogas plant could be sustained only recirculating the water without adding potable water in the process. In addition, the solid particles present in wastewater are mineralized in biodigesters, which will later be converted into biogas (Kulakov 2013; Volkov et al. 2014; Makisha 2016).

Despite the unavailability of data on the amount of water that is recycled and discarded in the biogas production process, is possible to conclude that greatest water consumption occurs in the stage of operation the biodigesters. It can be said that biogas production process itself does not have a high water footprint. The consumption of water during the process is difficult to calculate in numbers and the most significant is linked to different raw materials, especially those that go through cycles of cultivation, requiring large amounts of water for its development (Hoekstra et al. 2011; Zhao, Chen and Yang 2014).

The reuse of wastewater in productive processes emphasizes a great impasse, which is the shortage of food supply and drinking water. However the development of resource reuse technologies brings a solution to this environmental issue, promoting the expansion of industries with projects sustainable use of water reuse, even if the use of these technologies still do not have as many cases on a relative scale, study shows themselves promising (Lu et al. 2019). The digestate can be considered a by-product of anaerobic digestion due to the economic and energy value added to the process of universalization of sanitation (Felca et al. 2018). This biomass consists of solid and liquid portions that are possible to be separated for different applications; the two fractions are extremely rich in nutritional terms and have potential to be used in crop and soil fertilization (Esteves et al. 2019). An interesting point that comes close to the circular economy is that after going through separation processes the liquid from the digestate can be recirculated in the biogas production process as a dilutive agent of the organic load input the system (Figure 1) (Bacenetti et al. 2013; Esteves et al. 2019).

In this sense, different treatment forms are available in the literature aiming at the separation of the solid and liquid parts of the digestate. This separation can be done by pressing, centrifugation, among others. In these cases, the solid part presents the possibility of being incorporated to composting processes or fertilizers, provided that they are stored in tanks or decanters to reduce level of ammonia, which also has significant environmental advantages, since it reduces the volatilization of CH_4 to the atmosphere and the possibility of contamination of the soil, a positive consequence of the restriction of pathogens due to high temperature that the digestate is exposed (Tem Hoeve et al. 2014; Fantin et al. 2015; Esteves et al. 2019).

Within the biogas production cycle, environmental impacts are pointed out as reduced when compared to other biofuels processes. Taking into account factors such as definition and pre-treatment of the raw material, technology, operation, maintenance, feeding frequency among others has a direct influence consumption of resources, such as water and energy (Poeschl, Ward and Owende 2010; Uusitalo et al. 2013; Esteves et al. 2019). Investments in basic sanitation and in the efficiency of these systems, such as anaerobic digestion for energy generation, increase sustainability and present more environmentally viable strategies for the use of biogas (Felca et al. 2018).

Bioethanol

Regarding bioethanol, several studies have been carried out due to its low cost of production and the low emission of greenhouse gases (Chavez-Rodrigues and Nebra 2010). On the other hand, there is growing concern about the high amounts of water used in the process of obtaining this biofuel, as well as causing a social conflict regarding the withdrawal of water from water bodies for use in biofuel processing plants (Berndes 2002; Gerbens-Leenes et al. 2009; Chavez-Rodrigues and Nebra 2010; Fingerman et al. 2010).

Based on the production differences of each ethanol generation, the water consumption for each one varies, in relation to the raw material of each process and the technology applied for each industry (Zhang et al. 2010; Pina et al. 2017). In a 1G ethanol production plant from sugarcane with an integrated sugar production process, an average of 13 to 15 m^3/ton. sugarcane is used, four processes being responsible for 65% of consumption, the cane washing section (3 m^3/ton), the barometric condenser of evaporation (2.5 m^3/ton), cooling of fermentation vats (1.7 m^3/ton) and vacuum in the pans system (2.4 m^3/ton) (Chavez-Rodriguez et al. 2013; Pina et al. 2017). Using corn as the raw material for production, the average consumption is 10 liters of water per liter of ethanol produced, with 53% of consumption being required for cooling towers and 42% for steam generation for drying (Wu et al. 2009). Beet as the main source of raw material, the older industries consume on average 0.25 to 0.45 m^3/ton beet (Vaccari et al. 2005). In a mill with an integrated 1G and 2G ethanol production system from sugarcane, 2G production can consume 24 m^3/ton, where the raw material is pre-treated by a steam explosion (Mosqueira-Salazar et al. 2013). In the production of 2G ethanol from switchgrass, the consumption can be from 2 to 9 liters of water per liter of ethanol (Wu et al. 2009), with the cooling process is the largest water consumer, as well as the generation of effluents in condensation when the production process is thermochemical (Martín, Ahmetovic and Grossman 2011). In Taiwan were conducted surveys on the water consumption of an ethanol production plant 2G using bagasse sugarcane and rice straw. The process consisted of pretreatment by steam explosion catalyzed by sulfuric acid, followed by the process of saccharification and fermentation simultaneous. In this plant, the main operations with freshwater consumption were three: the steam boiler, the acid hydrolysis and the cleaning of the raw material (Chiu et al. 2015).

Before the studies of water reuse and recirculation strategies, it is necessary to understand the water quality required at each stage in order to replace fresh water, so that adjustments are made in the process or to avoid damages to equipment and the efficiency of the conversion into biofuel.

The treatment of the wastewater or even its reuse is important also for the chemical and physical characteristics. The characteristics vary depending on the type of effluent and on the process from which it originates. Furthermore, the process varies from one plant to another in relation to the type of raw material being used (Chavez-Rodrigues et al. 2013).

Vinasse is the largest amount of effluent generated in ethanol production processes. However, because it presents high-suspended solids, high BOD and low pH, direct reuse becomes difficult, requiring advanced treatment processes (Chavez-Rodrigues et al. 2013, Pina et al. 2017). Therefore, the vinasse will be disregarded in this section for strategies of reuse and recirculation.

In a biofuel production plant, water consumption can be reduced by increasing the use of measures such as process water reuse and recycling as well as process implementations and modifications using modern conversion technologies (Wu and Chiu 2011). The use of water can be minimized by various strategies such as increasing the recycle of process water, which can occur through the water vapor of the dryer, and recycling the condensate of the boiler, increasing the cycles in the cooling towers or recirculating the water cooling towers (Wu et al. 2009).

Recirculation of water in industrial processes producing 1G ethanol would be sufficient to reduce more than 90% of the extraction of water from natural sources of freshwater. Evaluated by Chavez-Rodrigues et al. (2013), in this proposal for these reuse circuits the processes that demand for a higher quality of water use the extraction by natural sources. The other processes are complemented from the recycled water in the industry, being able to reduce in the industries that use sugarcane as raw material, water consumption of 15 m^3/ton to 1.2 m^3/ton. Also, using water reuse streams the entire industry demand could potentially be supplied by evaluating the process of water quality, with condensates being the main potential sources of reuse, accounting for 43% of this potential. According to Chavez-Rodriguez et al. (2013), considering the reuse and recycling of water in ethanol production industries, water consumption would decrease by 15 m^3/ton to 0.4 m^3/ton. In this same context, Pina et al. (2017) evaluated the main demands of an ethanol production industry from sugarcane and listed the main potential sources of water for reuse within the industry, being filtration condensate filtration stage, evaporation system, condenser, a part of the washing water, the boiler purge and the water of the dehydration process, totaling 0,2 m^3/ton for the ethanol industry, and 0.7 m^3/ton for the industry with the integrated production process of sugar and ethanol. By effectively collecting water only to restore process losses and to meet high-quality demands such as imbibition and dilution of yeast, the effective water requirement would drastically reduce to approximately 1 m^3/ton.

In addition to the strategies of reuse and recirculation flow within the industrial process, thermal process integration is approached as a strategy to reduce freshwater consumption, mainly by reducing the demand in the cooling process in condensers, and by enabling the expansion of the scale and integration of 1G and 2G ethanol production processes (Dias et al. 2013; Oliveira et al. 2016; Pina et al. 2017). Pina et al. (2017) studied the thermal integration strategy, resulting in an effective reduction of water consumption of 24% and 13% for ethanol distillation plant and integrated sugar and ethanol production industry, respectively.

With the technological innovations, the raw materials of the biofuels production, mainly of the 1G tend to mechanization. And adaptations in the production process have been made, mainly in the stage of washing the raw material. The processes of washing the raw material are responsible for a high water consumption, for example, for cane sugar an average consumption of 3 m^3/ton and for corn of 0.5 m^3/ton. (Karuppiah et al. 2008; Martín, Ahmetovic and Grossman 2011; Chavez-Rodrigues et al. 2013). Currently, with the mechanization of the sugarcane harvest, dry cleaning processes, significantly reducing the water consumption (Chavez-Rodrigues et al. 2013), are carrying out the washing processes in the ethanol industries that use this raw material. In addition, when there is a washing process, the water can be recirculated after decantation processes, keeping the pH always basic (Karuppiah et al. 2008; Chavez-Rodrigues et al. 2013).

Another method would be to use ways that decrease water consumption in ethanol production. Mosqueira-Salazar et al. (2013), shows in his study in an integrated first and second generation bioethanol production plant that if the water circuits of the production process, such as cooling tower water, cooling water from the spray pond among others such as boiler water and boiler discharge are closed, water consumption significantly decreases compared to open circuit. In addition, another way to reduce water consumption is to reuse water in streams, the condensates in a bioethanol plant, for example, which are the second largest reuse stream after vinasse water (Chavez-Rodriguez et al. 2013).

It is possible to achieve the least need for fresh water extraction by reusing and recycling water using existing commercial technology and with additional capital investment (Wu et al. 2009). Newly built ethanol plants with efficient design and process integration can reduce the use of water. The development of a process design that optimizes water use should be encouraged from the outset (Wu and Chiu 2011). The challenge is to ensure the sustainability of the 1G ethanol business while attaining the 2G and 3G ethanol potential (Araújo 2016). From this, economic and environmental strategies are emerging in order to reduce the use of water in ethanol production, mainly aiming at the high water footprint of the process.

Recent studies discuss different strategies that are being evaluated as possible adaptations of traditional ethanol production plants. A technological strategy widely debated in recent studies is the substitution of freshwater for ethanol production by the use of seawater, mainly in coastal areas and arid zones (Greetham et al. 2018). The number of minerals present in seawater can dispense with the addition of nutrients that are required in alcoholic fermentation (Lin et al. 2011). Recently, studies have been focusing on alternative approaches to overcome the challenges of this strategy, mainly because of the microorganisms, since the majority of the yeasts used for bioethanol synthesis were isolated from terrestrial environments and cannot survive in a high concentration of salt and nutrients (Greetham et al. 2018). As well as the use of alternative marine biomasses, which generally have a different monomer spectrum of terrestrial plants, making it difficult to efficiently hydrolyze sugars by microorganisms (Jung et al. 2013).

Recently, Zaky et al. (2016) isolate 116 yeasts of which six showed greater capacity than the reference yeast of ethanol production processes, *Saccharomyces cerevisiae*, with the species being isolated: *S. cerevisiae*, *Candida tropicalis*, *Candida visvanathii*, *Wickerhamomyces anomalus*, Candida glabrata and P*ichia kudriavzevii*. Comparison of strains of terrestrial and isolated *S. cerevisiae* isolated from the sea has been studied, where the marine lineage presented higher potential yield for ethanol production in relation to terrestrial lineage (Saravanakumar, Senthilraja, Kathiresan 2013). The efficiency of conversion of sugars into ethanol of 66% has already been reported for yeast *Candida* sp. in high salt concentration using sugarcane hydrolyzate (Khambaty et al. 2013).

Marine microorganisms have also been used in enzymatic hydrolysis processes because of their ample capacity to produce enzymes such as amylases, proteases, cellulases, xylanases, and lipases, among others (Chi et al. 2016). Marine yeasts are still used as donors of genes for the development of new strains producing ethanol, aiming to increase the resistance of terrestrial strains and to increase the efficiency in the conversion of sugars (Zhang et al. 2010; Greetham et al. 2018). Thus, the use of seawater in the production of ethanol, from hydrolysis to the fermentation process, could improve the overall economy of the process, reducing the need for fresh water in a biorefinery. Therefore, the development of bioethanol strategies based on seawater can certainly have a strong impact on overcoming freshwater and energy crises (Zaky et al. 2014).

The reduction of the demand for fresh water in the process of cellulosic ethanol production was studied through the substitution of treated wastewater. The most consumed stages are pre-treatment, biomass washing, hydrolysis, and fermentation. Ranchandran et al. (2013) evaluated the substitution in the hydrolysis and fermentation process using commercial cellulase enzyme and *Saccharomyces cerevisiae*. The concentrations of glucose and ethanol were similar for different proportions of the effluent use, demonstrating the feasibility of replacing freshwater with wastewater. Nevertheless, the authors affirm that there is a need for extensive studies to adopt this practice on an industrial scale.

In relation to 3G ethanol, there is still a large-scale production problem, mainly because different algal species require different approaches to cultivation; in addition, the crop requires specific reaction conditions for effective growth, including intensities of light, temperature, gas exchange and nutrient composition (Kadir et al. 2018). In addition, in order to improve and adapt certain species to different conditions, many studies bring approaches to strategies to reduce water consumption through wastewater substitution (Ge, Madill and Champagne 2018).

As seen in this section, biogas and bioethanol are obtained from complex processes in which the use of water is imminent, most often in good quality to generate a valuable product. Some strategies for reducing the water footprint in both processes are already seen in real plants, but others still require further studies, which in turn will take longer to effectively been adopted making the production mills continue to operate according to as they came, generating a large amount of water at the end of the productive cycle. Some works contained in the literature reinforce the great interest to provide reuse of generated wastewater. For this, steps of treatments of this material are necessary. The following will be raised the most common types of treatments, among which are those related to the capacity for reuse of water in the production plant itself or for the purpose of promoting other uses.

Treatments to Ensure the Recycling and/or Reuse of Wastewater

Biogas

The singular approach of treat or recycle biogas effluent or sludge just with the purpose of remove Chemical Oxygen Demand (COD), Nitrate (N) and Phosphorus (P) is letting space to new simple technologies which embrace the contaminants removal and the generation of by-products capable of be inserted in industrial processes. This new circular economy based approach is been studied in three main areas, which are biogas wastewater use as a fertilizer, as a nutrient source to cultivate algae and as a biogas enhancer.

Fertilizer Purpose

The digestate, a liquid portion obtained from anaerobic decomposition in biogas plants, is considered rich in mineral elements such as nitrogen, phosphorous, potassium, among others. These elements are essential in agricultural crops, where the soil has to be fertile. Several studies already showed that digestate application for soil fertilization purpose improves plants quality and their immunity to pathogenic agents (Kouřimská, Poustková and Babička 2012; Chiew, Spångberg and Baky 2015; Koszel and Lorencowicz 2015).

For a successful application as a soil fertilizer, the nutrients have to be concentrated in some cases and the liquid digestate need pretreatment to decrease its mass and improve the fertilizing power. Besides that, the treatments reduce transportation costs by reducing the final product volume. Combining techniques of solid-liquid separation and digestate liquid treatments, the result can be a fertilizer with optimal composition and the water portion can be separated and recirculate to biogas plant (Hjorth et al. 2010; Rehl and Müller 2011; Chiumenti et al. 2013; Tampio, Marttinen and Rintala 2016). A circular economy approach can be designed in this digestate utilization for fertilization purpose since the agricultural culture has its growth improved with fertilizer and the plant waste material serve as a substrate to a biogas plant, which provides the digestate to generate the fertilizer.

According to Esteves et al. (2019), the solid-liquid separation provides two fractions with great potential as fertilizer to be applied on crops. For this separation occurs, some treatment technologies can be used and provides good nutrient distribution. The most cited treatments are ammonia stripping, centrifugation, evaporation, separation with membrane, precipitation, decantation as well as the combination of more than one technique mentioned (Ek et al. 2006; Chiumenti et al. 2013; Ledda et al. 2013; Ten Hoeve et al. 2014; Tampio, Marttinen and Rintala 2016; Esteves et al. 2019). The technology choice, in most cases, are made based on the greatest fertilizer potential obtained, but Tampio, Marttinen, and Rintala (2016) brings the importance of considering factors such as process inputs and outputs, energy balance, mass flows and others factors, known as the study of the chain life cycle assessment, which considers environmental and economic aspects.

After receiving the appropriate treatment, digestate must be temporarily stored in order to reduce ammonia levels and avoid plantation burn. The storage can be made in wells, tanks or ponds (Esteves et al. 2019). When the fertilizer is ready to apply in the soil surface, it has to be chosen the best method to administer considering the digestate treatment technique, the agricultural crop type and the time of fertilization (Koszel and Lorencowicz 2015).

According to Xia and Murphy (2016), the demand for digestate use as a fertilizer is huge but, in some countries, there is the barrier of obtaining authorization to use this product because of the difficulty to control the efficiency of digestate treatment and application methods, especially in concern of the presence of the heavy metals in some wastewater from anaerobic digestion.

Algae Cultivation Purpose

Algae are photosynthetic organisms having a great potential to be applied in biofuel, biochemical and nutrition areas, generating valuable products for commercial application. However, the algae cultivation scale-up have problems, such as the nutrient cost and energy input requirement (Cheng et al. 2015; Zhu 2015; Xia and Murphy 2016; Zuliani et al. 2016).

To solve part of the problems, the combination of treat liquid digestate from biogas anaerobic reactor and cultivate algae have been gained space in researches. The liquid digestate is rich in nitrogen and phosphorus, the primary nutrients for algae growth. In addition, the biogas wastewater systems with algae application have the aeration facilitated under no or less mechanical procedures (Wang et al. 2014; Xia and Murphy 2016; Cai et al. 2019). Because of these facts, the algae cultivation in biogas plants is beneficial both for algae growth and for anaerobic digester effluents treatment, as shown in Table 1.

Franchino et al. (2013) selected three strains of *Chlorophyceae* (*Neochloris oleoabundans*, *Chlorella vulgaris,* and *Scenedesmus obliquus*), already known for their productivity, with nutrient removal purpose from the effluent of a pilot anaerobic digester fueled with cattle slurry and raw cheese way. As can be seen in Table 1, the digestate COD concentration was 3290 mg/L, a 1:10 dilution factor and the optimal concentration of reactor fed was 35% of cattle slurry and 65% of raw cheese way, is possible to obtain algae biomass between 1400-1610 mg/L. The authors tested others digestate dilutions, but the COD concentration reported in Table 1 was the most promising, being able to result in great N and P removal, above 83.7% and 96.0%, respectively, by all three strains.

In Wang et al. (2015) work (Table 1), it is also possible to compare microalgae species growth and COD, N and P removal under the same COD concentration in the digestate (789.13 mg/L). Although lower COD quantity was available to microalgae, *S. obliquus* was capable to have a similar biomass growth as Franchino et al. (2013) study, but the N and P removal was lower. The best COD removal (68.11%) was assigned to *Nitzschia palea*, the microalgae species capable to remove the greatest values of Nitrogen (59.08%) and Phosphorous (60.03%) in this study. For all microalgae species, Wang et al. (2015) reported that moderate photoperiod (14 h light and 10 h dark) was the best treatment to have greatest growth rates, suggesting that this parameter is a limiting factor and provide a simple and effective operational strategy to increase the process efficiency.

Table 1. Removal of Chemical Oxygen Demand (COD), Nitrogen (N) and Phosphorous (P) from anaerobic digester effluents with algae cultivation purpose

Reactor fed	Algae cultivated	Digestate concentration (mg COD/L)	Algae biomass (mg/L)	COD removal (%)	N removal (%)	P removal (%)	Reference
Cattle slurry (35%) and raw cheese whey (65%)	*Neochloris oleoabundans*	3290	1400	-	99.9	96.9	Franchino et al. (2013)
	Chlorella vulgaris		1610		99.9	96.0	
	Scenedesmus obliquus		1540		83.7	96.1	
Not reported	*Scenedesmus obliquus*	789.13 ± 19.88	1057	60.39 ± 6.12	56.71 ± 5.17	52.97 ± 3.75	Wang et al. (2015)
	Selenastrum capricornutum		658	55.48 ± 5.46	45.72 ± 4.36	40.95 ± 5.19	
	Nitzschia palea		749	68.11 ± 6.51	59.08 ± 5.21	60.03 ± 5.38	
	Anabaena spiroides		553	61.68 ± 3.95	58.35 ± 5.39	36.58 ± 2.91	
	Chlorella vulgaris		973	64.76 ± 5.37	55.67 ± 4.93	53.84 ± 4.32	
Piggery wastewater pretreated	*Scenedesmus obliquus*	3200	1691.69	65.06 ± 7.18	63.37 ± 5.16	71.61 ± 6.83	Xu et al. (2015)
		2200	1844.15	73.41 ± 6.12	69.61 ± 6.28	70.09 ± 5.87	
		1600	2178.96	75.29 ± 5.89	74.63 ± 6.94	81.73 ± 7.28	
		1200	1577.17	72.29 ± 6.75	62.54 ± 4.97	88.79 ± 6.24	
		800	1207.71	63.02 ± 4.86	58.39 ± 5.13	85.18 ± 8.96	
		400	870.66	61.58 ± 5.71	59.07 ± 5.62	88.42 ± 7.65	
Seafood processing wastewater	*Chlorella* sp.	50[a]	184	-	85.45	8.36	Jehlee, Khongkliang and O-Thong (2017)
		60[a]	121		55.01	6.93	
		70[a]	61		50.41	5.58	
		80[a]	48		26.67	4.66	
		90[a]	39		21.91	4.58	
		100[a]	32		9.07	4.07	

Reactor fed	Algae cultivated	Digestate concentration (mg COD/L)	Algae biomass (mg/L)	COD removal (%)	N removal (%)	P removal (%)	Reference
Swine wastewater pretreated	*Chlorella vulgaris* with fungi	1041.27 ± 28.76	3017	86.08 ± 6.27	85.69 ± 6.34	86.17 ± 6.13	Wang et al. 2017
	Chlorella vulgaris with activated sludge		3003	84.28 ± 4.73	84.17 ± 5.54	83.79 ± 5.83	
Piggery wastewater	*Chlorella vulgaris*	1200 and 1600[b]	860	74	72	78	Gao et al. 2018
	Chlorella vulgaris with fungi		940	78	78	86	
	Chlorella vulgaris with activated sludge		1020	80	80	90	
	Scenedesmus obliquus		890	73	75	80	
	Scenedesmus obliquus with fungi		960	77	80	86	
	Scenedesmus obliquus with activated sludge		990	77	80	90	
Granular activated carbon-pretreated desizing textile wastewater	*Scenedesmus* sp.	3000	1410	51.6 ± 5.9	-	-	Nguyen, Lin and Lay (2019)
		5000	1790	69.1 ± 5.3			
		7000	320	2.4 ± 1.6			

- Not quantified.

[a] In % of biogas effluent dilution (v/v).

[b] There was no significant difference between COD concentrations and results ($p > 0.05$).

As well as Franchino et al. (2013) and Wang et al. (2015) studied the *S. obliquus* behavior growing in biogas effluent, Xu et al. (2015) research submitted this algae specie to different digestate concentrations from a reactor supplied with piggery wastewater and analyzed the growth and nutrient removal responses (Table 1). Before submitting the microalgae to growth in effluent presence, the piggery anaerobic digester wastewater was pretreated through sedimentation and filtration processes, and the supernatant was autoclaved and applied to the treatment process with *S. obliquus*. As shown in Table 1, there is a significant difference between COD, N and P removal according to digestate dilution. In 7 days batch culture, the microalgae reached until 2178.96 mg/L of biomass under 1600 mg COD/L condition, having good performance in nutrients removal. But, comparing this study results to Franchino et al. (2013), with a similar digestate concentration and the same microalgae species (see Table 1), it is possible to infer that the reactor fed parameter have a considerable influence in N and P removal.

Other microalgae genera widely studied for biogas wastewater treatment purpose is *Chlorella* sp., as reported by Jehlee, Khongkliang, and O-Thong (2017). In their research, the growth rate and nutrient removal from a biogas effluent of seafood processing were analyzed, as can be seen in Table 1. The effluent was diluted in six different concentrations and in 6 days, it was found that the highest growth of microalgae occurs in 50% dilution (reaching to 184 mg/L of algae biomass). In addition, this dilution had the highest N and P removal values, being 85.45% and 8.36%, respectively. Comparing the N removal with others works in Table 1; it is possible to see the great *Chlorella* sp. performance in Jehlee, Khongkliang, and O-Thong (2017) even with lower growth rates. In the other hand, the P removal was not satisfactory, being possible to assume the connection between algae growth and Phosphorous consumption.

In order to improve the nutrition assimilation and its removal from swine wastewater by *Chlorella vulgaris*, Wang et al. (2017) proposed in their work the co-cultivation of this species with fungi *Ganoderma lucidum* and with nitrifying-denitrifying activated sludge obtained from a wastewater treatment plant. The biogas slurry from anaerobically digested swine wastewater was pretreated, before application in microalgae growth, through ultraviolet sterilizer and filtration in order to remove the bulky grain. According to Wang et al. (2017), the wastewater treatment efficiency was strongly dependent on microalgae growth rate, idea proposed in the preceding paragraph and that is possible to confirm observing Table 1, where a low difference between algae biomass resulted in greater results in COD, N and P removal (around 86%) for *C. vulgaris* cultivated with fungi. The study also reported the importance of treatment with appropriate light wavelengths, having an improvement in nutrient removal efficiency when red to blue light ratio was fixed in 5:5.

Great results when *C. vulgaris* is co-cultivate also appears in Gao et al. (2018) study, where the authors compare the performance between two algae specie (*C. vulgaris* and *S. obliquus*) mono-cultivation and co-cultivation with fungi *G. lucidum* and with activated sludge collected from a wastewater treatment plant. The three configurations of cultivation were inoculated with piggery wastewater and the results obtained with 1200 and 1600 mg COD/L in the digestate is shown in Table 1. The *C. vulgaris* biomass results in co-cultivation configuration can be compared to Wang et al. (2017), where the main difference between studies is the wastewater pretreatment before microalgae inoculation, which provides to the *C. vulgaris* a greatest growth rate. In addition, the lack of pretreatment made the COD and N removal reach just until 80% in Gao et al. (2018) work for algae cultivated with activated sludge. In contrast, this configuration process had the greatest results on P removal (until 90%) comparing to wastewater with pretreatment in Wang et al. (2017).

To *S. obliquus* cultivation, the configuration with activated sludge also gave the best results in nutrient removal from piggery wastewater (see Table 1), removing 77% of COD. In Nguyen, Lin and Lay (2019) research, it was possible to obtain similar results in COD removal (69.1%) with *Scenedesmus* sp. mono-cultivation in granular activated carbon-pretreated desizing textile wastewater in a higher digestate concentration (5000 mg COD/L). The digested effluent was pretreated through centrifugation procedure and its supernatant was used to *Scenedesmus* sp. cultivation in biogas CO_2 presence in a concentration of 22%. After ten days of algae cultivation, it was possible to compare growth rate with different digestate concentration, having a COD removal and biomass peak with intermediaries' values and decreasing with the higher value, as shown in Table 1, proving the strong relation between initial COD concentration, algae growth, and nutrient removal.

According to Xia and Murphy (2016), the nutrient removal efficiency has great variations depending on algal species, biogas effluent characteristics, operational parameters, among others factors, which can be verified in the study's results summarized in Table 2. With this, the advantages in treat the wastewater from anaerobic digesters and also obtain algae with a large range of applications can not be denied, having great COD, N and P removal efficiency and being able to also remove heavy metals, according to Cheng et al. (2015).

Biogas Upgrading Purpose

The circular economy approach is gaining more space in biogas production, especially when the process effluent gives the perspective to upgrade the product. As mentioned in a previous section, biogas is a biofuel obtained through anaerobic digestion of a several ranges of wastes, serving as a substrate to the microorganisms engaged in the process of converting them into CH_4, CO_2 and digestate, the main compounds formed.

Although the anaerobic digestion process embraces the production of a gas portion formed by methane and carbon dioxide, the industrial interests in biogas spin around CH_4. The highest are CH_4 quantity, the purer is consider the biogas. Therefore, the term biogas upgrade refers to remove the other compounds and let the major portion of methane, as it is possible.

The treatments to remove nutrients from liquid digestate was cited in the sections above, but some of the same processes can integrate the COD, N and P removal with CO_2 removal. The algae metabolism have the capability to assimilate this gas during photosynthesis, but it must be sorted according to their growth and carbon dioxide fixation tolerance, as report Wang et al. (2015). In their research, the initial CH_4 concentration was upgraded from 61.38% (v/v) to between 85-95% receiving treatment with *C. vulgaris*, *S. obliquus*, *S. capricornutum*, *N. palea* and *A. spiroides*.

Xu et al. (2015) also registered the CO_2 consumption during *S. obliquus* photosynthesis and growth, with a CO_2 removal ranged from 54.26 to 78.81%, confirming this microalgae species ability of upgrade biogas in the same time that is removing nutrients from liquid wastewater. According to the authors, the CO_2 reduction in biogas is stimulated during the nutrient loading, mainly because microalgae in its growth process consume both substances. Besides that, Xu et al. (2015) reported the CO_2-concentrating mechanism, developed by *S. obliquus* in order to adapt its metabolism to changes in CO_2 concentration.

In Zeng et al. (2018) research, the objective was to remove H2S from biogas using synthetic wastewater and biogas slurry. Simultaneous biogas desulfurization and wastewater denitrification system was successfully developed in this study, being capable of remove 84.7% of H2S and 60.9% of NOx^--N and providing a feasible route to biogas desulfurization and, consequently, upgrade through biogas slurry use as a nutrient.

Bioethanol

The existing treatment technologies for recycling or reuse of wastewater are numerous. These involve a number of benefits such as the protection of human health as well as ecosystems. In addition, the treatment of wastewater makes it possible to return, in good conditions, to the natural reserve of water in the world (Ding 2017). Thus, the problem related to the availability of water, ensuring that there is no contamination of water bodies and ensuring a greater supply of drinking water (Gupta et al. 2012).

Generally, the type of treatment to be applied involves three steps: primary, secondary and tertiary treatment and its application are related to the type of contaminant involved and the desired use for water after treatment (Sonune and Ghate 2004; Gupta et al. 2012).

Currently, the main methodologies for treatment are based on the traditional methods that comprise anaerobic and aerobic digestion with the aim of reducing the chemical oxygen demand (COD) of the wastewater, being within the parameters allowed by the legislation (Zhang et al. 2010). However, this technique may have low methane yields due to the presence of recalcitrant organic compounds (Chen, Cheng and Creamer 2008).

On the other hand, existing methodologies are being thought in the last years mainly aiming to avoid the expense of resources and environmental damages. Coagulation (Kumar et al. 1997), evaporation (Larsson et al. 1997) and electrolysis (Yavuz 2007) are some examples of methodologies suggested for the treatment of wastewater from bioethanol production industries. Another factor is that, if properly treated, they can be returned to the fermentation process, enabling the recycling and reuse of wastewater (Zhang et al. 2010).

Table 2. Characteristics of various types of distillery wastewater

Parameter	Distillery wastewater	Wine distillery wastewater	Vinasse	Raw spent wash	Lees stillage	Molasses wastewater
BOD_5 (g/L)	30	0.21-8.0	42.23	-	20	-
COD_T (mg/L)	100-120	3.1-40		37.5	-	80.5
COD_S (mg/L)	-	7.6-16	97.5	-	-	-
TOC (mg/L)	-	2.5-6.0	36.28	-	-	-
pH	3.0-4.1	3.53-5.4	4.4	4-5	3.8	5.2
EC	346	-	-	2530	-	-
Alkalinity (meq/L)	-	30.8-62.4	-	2	9.86	6000
Phenol (mg/L)	-	29-474	477	-	-	45.0
VFA (g/L)	1.6	1.01-6	-	-	0.248	8.5
VS (g/L)	50	7.340-25.4	-	-	-	79
VSS (g/L)	2.8	1.2-2.8	-	-	0.086	2.5
TDS (mg/L)	-	-	51.500	-	-	51.500
TS (g/L)	51.5-100	11.4-32	1.5-3.7	2.82	68	109
TSS (g/L)	-	2.4-5.0	-	-	-	-
MS (g/L)	-	6.6	-	-	-	30
MSS (g/L)	-	900	100	-	-	1100
TN (g/L)	-	0.1-64	-	2.02	1.53	1.8
$NH4^+$ (mg/L)	-	140	-	125-400	45.1	-
$NO3^-$ (mg/L)	4900	-	-	-	-	-
TP (g/L)	-	0.24-65.7	-	0.24	4.28	-
PO_4^{3-} (mg/L)	-	130-350	-	139	-	-
Iron	0.06	0.05-0.075	-	-	-	0.028

- Not quantified.

Source: Adapted from Bustamante et al. 2005; Nataraj, Hosamani and Aminabhavi, 2006; Melamane, Strong and Burgess, 2007; Yadav and Chandra, 2012; Prajapati and Chaudhari, 2015.

Ethanol production distilleries generate high volumes of wastewater characterized by dark staining, acid pH, high BOD (Biochemical Oxygen Demand) concentration and COD (Chemical Oxygen Demand), dissolved solids, total solids, total suspended solids, and high concentrations of nitrogen, potassium, phosphorus, calcium, and sulfate (Table 2). This composition is due to the presence of organic matter in the form of proteins, sugars, polysaccharides, lignin, melanoidins, and waxes, besides the presence of recalcitrant pollutants (Nataraj, Hosamani and Aminabhavi, 2006).

Because of the characteristics of wastewater from alcohol distilleries, the application may be limited. The presence of some compounds in these effluents may even hinder biodegradation. However, in order to reduce water consumption in the process of obtaining biofuels, it is suggested that these wastewater be treated or reused, returning to the process for steam generation, cooling tower water, equipment others (Singh et al. 2018).

Microfiltration (MF) is a membrane separation process that could be useful in the retention of suspended solids but is unable to remove organic and dissolved color. However, an action combined with biological or chemical treatment would be able to meet the quality requirement of the treated effluent (Basu et al. 2015).

In the study by Zhang et al. (2010), a treatment that consisted of the recycling of wastewater from the bioethanol production process was proposed. In this case, a treatment was performed that consisted of the anaerobic treatment with two stages, with the concomitant production of biogas.

The liquid portion resulting from anaerobic digestion was recycled for use in the fermentation to obtain ethanol. Thus, the anaerobic digestion step was effective in removing most of the impurities present in the wastewater, while producing significant amounts of biogas, which can be applied in the generation of energy. In this way, the generation of wastewater becomes very low, in addition to significantly reducing energy consumption (Zhang et al. 2010).

However, this process is not so simple due to the high concentrations of organic matter present in the wastewater resulting from the bioethanol production process (Yang et al. 2015). An alternative is the application of effluent generated in different sectors, such as organic production (Yang et al. 2015), which still has low added value, methane gas generation (Rodriguez et al. 2013), which requires long retention times and the generation of electricity (Quan et al. 2014) that has a limitation due to the high cost (Xin et al. 2017).

On the other hand, the use of wastewater from ethanol distilleries to obtain hydrogen has been studied. In the study developed, the percentage hydrogen yield obtained was 40% and hydrogen yield of 125 mL H_2/g COD was removed, using a sequential batch anaerobic reactor (Searmsirimongkol et al. 2011; Intanoo et al. 2014). Lazaro, Varesche and Silva (2015), made use of fermentation and phototrophic system to produce hydrogen, obtaining more than 70% of COD removal and hydrogen yield of 123 mL H_2/g COD removed, making use of an anaerobic fluidized bed reactor. Studies based on the production of hydrogen from wastewater shows itself as an attractive alternative due to the valuation of these residues, transforming them into products with high benefit (Searmsirimongkol et al. 2011).

Regarding the characteristics of the wastewater from ethanol production, these vary according to the substrate used. For example, using cassava as the substrate for ethanol production, large quantities of effluents mainly containing acids and carbohydrates, as well as organic compounds that can be soluble and insoluble such as starches, sugars, proteins, and cellulose (Quan et al. 2014). Consequently, the effluent has a high COD, BOD and total solids (Hien et al. 1999; Kaewkannetra et al. 2009).

In the case of sugarcane processes among the stages of production are those involving the crystallizers, where the liquid from the process passes by vacuum boiling to form crystals and condensers responsible for receiving all the condensates generated in the previous steps, except for the vapor that returns to the cogeneration system automatically. These condensates remain in storage tanks, allowing them to return to the process depending on their state (Chavez-Rodrigues et al. 2013).

Other by-products resulting from the process is vinasse, followed by condensate. However, vinasse already has relevant applications such as fertigation technique, which can also be considered practice for reducing water consumption. The reuse of the vinasse becomes impracticable due to the fact that it has a high concentration of suspended solids, high BOD values, and low pH. In the same way as the water from the sugar cane washing, requiring advanced treatments for reuse. As a result, condensate wastewater can be said to be the main form of reuse, representing about 43% of the water reuse capacity. (Chavez-Rodrigues et al. 2013).

In the case of the boiler, the treatment step is usually more complete because the pH remains high and there is no dissolved oxygen during the treatment (Kalakodimi et al. 2009). In this way, the use of demineralization or deionization techniques is recommended, causing the present íons to be removed (Chavez-Rodrigues et al. 2013). In addition to these, ultrafiltration and reverse osmosis techniques can also be used for the same purpose (Anis, Hashaiken and Hilal 2019).

The temperature of the water must be taken into account because, in the case of yeasts, it is recommended to use cold water and in good cleaning conditions so that there is no increase in the cooling system (Rawat 2015). Therefore, there should be a preliminary treatment when using the water from the condensate.

On the other hand, for the dilution of the molasses from the process, and for the washing of the sugar present in the centrifuges, the usual temperature is 80°C. Considering the washing of the cake/dough, it is necessary that the water present a good quality condition, as it will be part of the final product. The temperature range around 80°C ensures that there is no proliferation by bacteria and still prevents impermeability (Junzhang et al. 2014).

In the case of cooling towers, a chemical treatment is necessary to allow the adjustment of pH, which can be done with the addition of lime or chemical products, and bacterial control so that future problems do not occur, such as corrosion, obstruction and the reduction of heat transfer Chavez-Rodrigues et al. 2013). Condensate water can be reused in cooling towers; if the BOD is in acceptable condition, otherwise the use of chemical treatment is required (Wan, Sadhukhan and Ng 2015).

The practice of washing sugarcane, in turn, tends to decrease progressively due to the adoption of mechanized techniques. In addition, when the sugar cane is washed, considerable losses of sugar can occur. Moreover, when it is carried out, the water goes through decanting and recirculated to the process with basic pH so that degradation does not occur. Finally, water that has a low hardness, ie less than 80 mg $CaCO_3$/L should be used for cleaning equipment as well as general floor cleaning (Chavez-Rodrigues et al. 2013).

Usually, the alcoholic fermentation process generates residues that are treated by solid-liquid separation and are then referred for anaerobic digestion. Finally, the effluent generated in this last step is treated by conventional physical or chemical methods until reaching the standards established by the legislation for disposal (Kim et al. 1997).

Some studies have been carried out with the objective of evaluating the reuse potential of the waste generated during the process of obtaining the ethanol. That is the case of anaerobic digestion of vinasse, the main resulting residue. This can be sent to obtain biogas, being an alternative to the treatment of vinasse (Lee et al. 2011). In addition, after the anaerobic digestion step, the stillage can be returned to recirculation and reused as process water. Thus, the quantity of fresh water to be used is significantly reduced (Alkan-Ozkaynak and Karthikeyan 2011). Thus, anaerobic digestion offers benefits such as a reduction in water consumption and as a way to recover energy with the production of methane gas (Yang et al. 2016).

Another way to reduce water consumption in this process is through changes in fermentation processes. Shojaosadati (1996), evaluated the reduction of water using recycled biomass and vinasse, where recycling of 15% to 70% of vinasse from another fermentation process occurred, as well as a 13% to 47% decrease in water consumption and vinasse volume.

The vinasse has numerous applications and can be used together with the cooking and fermentation processes, which help in reducing its volume, as well as reducing the amount of water, as well as the chemical products to be used (Yang et al. 2016).

Another potential use of wastewaters would be for crop irrigation. Coupled with population growth, is the demand for more good quality food, which drives the agricultural sector. Thus, it is necessary to maintain irrigation of plantations with water of good quality and in sufficient quantity (Vanham et al. 2018). On the other hand, this is the sector that most demands quantity of water in good conditions, a factor that aggravates the water scarcity (Liu, Liu and Yang 2016). Therefore, the reuse of wastewater from processes such as the ethanol production process is a very attractive alternative to supply demand and consequently conservation of water resources (Matsumura and Mierzwa 2008; Vergine et al. 2017).

The cultivation of microalgae is also predicted in wastewater generated inherent to the production of bioethanol that, in addition to reducing the organic load present in these effluents, still use it as a source of nutrients for its growth, and consequently, biogas generation (Rawat et al., 2011). This process is considered ecologically correct since the biomass produced is reused, allowing the recycling of nutrients (Olguín 2003).

The uses of water in the industrial field have received great attention because of the high water footprint and because of the low quality that it is returned to the end of the productive cycle. Now there is the great problem of the water crisis in several parts of the globe that has come producing great concern about the maintenance of this resource in the future. Biogas and bioethanol are strongly dependent on water directly or indirectly in their raw material conversion processes, causing the generation of wastewater with high polluting potential due to the high organic load still present. As a result, this chapter sought to cover in the literature the strategies that are possible to be applied to reduce the consumption and reuse of water in the plant itself. Often the acceptance by these strategies is limiting because they generate high costs causing industries to opt for later treatments that guarantee the greater quality of the final effluent. In this way, the most used technologies for the treatment of effluents generated with a view to the reuse of this material were also addressed, as well as to develop the concepts of the circular economy within a biogas and bioethanol plant and promoting greater sustainability of the system.

REFERENCES

Al Seadi, T., Rutz, D., Prassl, H., Köttner, M., Finsterwalder, T., Volk, S. and Janssen, R. (2008). *Biogas Handbook*. Ed: Al Seadi, T. University of Southern Denmark Esbjerg, Niels Bohrs.

Alkan-Ozkaynak, A. and Karthikeyan, K. G. (2011). Anaerobic digestion of thin stillage for energy recovery and water reuse in corn-ethanol plants. *Bioresource Technology*, 102: 9891-9896.

Álvarez, J. A., Armstrong, E., Presas, J. and Gómez, M. S. (2004). Performance of a UASB-digester system treating domestic wastewater. *Environmental Technology*, 5:1189-1199.

Angelidaki, I. and Ellegaard, L. (2003). Codigestion of Manure and Organic Wastes in Centralized Biogas Plants. *Applied Biochemistry and Biotechnology*, 109: 95-105.

Anis, S. F., Hashaikeh, R. and Hilal, N. (2019). Reverse osmosis pretreatment technologies and future trends: A comprehensive review. *Desaliation*, 452: 159-195.

Araújo, W. A. (2016). "Ethanol industry: surpassing uncertainties and looking forward." In: *Global Bioethanol:* Evolution, risks, and uncertainties, edited by Sergio L. M. Salles-Filho, Luís A. B. Cortez, José M. F. J. da Silveira, Sergio C. Trindade, Maria G. D. Fonseca, 1-33. Massachusetts: Academic press.

Azhar, S. H. M., Abdulla, R., Jambo, S. A., Marbawi, H., Gansau, J. A., Faik, A. A. M. and Rodrigues, K. F. (2017). Yeast in sustainable bioethanol production: a review. *Biochemistry and Biophysics Reports*, 10: 52-61.

Bacenetti, J., Negri, M., Fiala, M. and Gonzalez-García, S. (2013). Anaerobic digestion of different feedstocks: impact on energetic and environmental balances of biogas process. *Science of the Total Environmental*, 463:541-551.

Balakumar, S. and Arasaratnam, V. (2012). Osmo-, thermo- and ethanol-tolerances of Saccharomyces cerevisiae S_1. *Brazilian Journal of Microbiology*, 2011: 157-166.

Bansal, V., Tumwesige, V. and Smith, J. U. (2016). Water for small-scale biogás digesters in sub-Saharan Africa. *Global Change Biology Bioenergy*, 9: 339-357.

Basu, S., Mukherjee, S., Kaushik, A., Batra, V. S. and Balakrishnan, M. (2015). Integrated treatment of molasses destillery wastewater using microfiltration (MF). *Journal of Environmental Management,* 158: 55-60.

Bechara, R., Gomez, A., Saint-Antonin, V., Schweitzer, J., Maréchal, F. and Ensinas, A. (2018). Review of design works for the conversion of sugarcane to first and second-generation ethanol and electricity. *Renewable and Sustainable Energy Reviews*, 91: 152-164.

Berndes, G. (2002). Bioenergy and water - the implications of large-scale bioenergy production for water use and supply. *Global Environmental Change*, 12: 253-271.

Bixio, D., Thoeye, C., De Koning, J., Joksimovic, D., Savic, D., Wintgens, T. and Melin, T. (2006). Wastewater reuse in Europe. *Desalination, Integrated Concepts in Water Recycling* 187: 89–101.

Bouallagui, H., Lahdheb, H., Romdan, E. B., Rachdi, B. and Hamdi, M. (2009). Improvement of fruit and vegetable waste anaerobic digestion performance and stability with co-substrates addition. *Journal of Environmental Management*, 90: 1844-1849.

Cai, W., Zhao, Z., Li, D., Lei, Z., Zhang, Z. and Lee, D-J. (2019). Algae granulation for nutrients uptake and algae harvesting during wastewater treatment. *Chemosphere*, 214: 55-59.

Cardona, C. A., and Sánchez, O. J. (2007). Fuel ethanol production: Process design trends and integration opportunities. *Bioresource Technology*, 98: 2415-2457.

Cardozo, N. P., Bordonal, R. O. and La Scala Jr, N. (2016). Greenhouse gas emission estimate in sugarcane irrigation in Brazil: is it possible to reduce it, and still increase crop yield? *Journal of Cleaner Production,* 112: 3988-3997.

Carrillo- Nieves, D., Alanís, M. J. R., Quiroz, R. C., Ruiz, H. A., Iqbal, H., M. N. and Parra-Saldívar, R. (2019). Current status and future trends of bioethanol production from agroindustrial wastes in Mexico. *Renewable and Sustainable Energy Reviews*, 102: 63-74.

Castellani, B., Rinaldi, S., Bonamente, E., Nicolini, A., Rossi, F. and Cotana, F. (2017). Carbon and energy footprint of the hydrate-based biogas upgrading process integrated with CO_2 valorization. *Science of the Total Environment*, 615: 404-411.

Chandler, J. A., Jewel, W. J., Gossett, J. M., Van Soest, P. J. and Robertson, J. B. (1980). Predicting methane fermentation biodegradability. *Biotechnology and Bioengineering Symposium,* 10: 93-107.

Chavez-Rodriguez, M. F. and Nebra, S. A. (2010). Assessing GHG emissions, ecological footprint, and water linkage for different fuels. *Environmental Science Technology*, 44: 9252-9257.

Chavez-Rodriguez, M. F., Mosqueira-Salazar, K. J., Ensinas, A. V. and Nebra, S. A. (2013). Water reuse and recycling according to stream qualities in sugar-ethanol plants. *Energy for Sustainable Development,* 17: 546-554.

Chen, S., Chen, B., and Song, D. (2012). Life-cycle energy production and emissions mitigation by comprehensive biogas-digestate utilization. *Bioresource Technology*, 114: 357-364.

Chen, Y., Cheng, J. J., and Creamer, K. S. (2008). Inhibition of anaerobic digestion process: A review. *Bioresource Technology*, 99: 4044-4064.

Chen, Y., Zhang, T., Yang, H. and Peng, J. (2016). Study on energy and economic benefits of converting a combined heating and power system to a tri-generation system for sewage treatment plants in subtropical area. *Applied Thermal Engineering*, 94: 24-39.

Cheng, J., Xu, J., Huang, Y., Li, Y., Zhou, J. and Cen, K. (2015) Growth optimization of microalga mutant at high CO2 concentration to purify undiluted anaerobic digestion effluent of swine manure. *Bioresource Technology*, 177: 240-246.

Chi, Z., Liu, G., Lu, Y., Jiang, H. and Chi, Z. (2016). Bio-products produced by marine yeast and their potential applications. *Bioresource Technology*, 202: 244-252.

Chiew, Y. L., Spångberg, J. and Baky, A. (2015). Environmental impact of recycling digested food waste as a fertilizer in agriculture – A case study. *Resources, Conservation and Recycling*, 95: 1-14.

Chiu, C. C., Shiang, W., Lin, C. J., Wang, C. and Chang, D. (2015). Water footprint analysis of second-generation bioethanol in Taiwan. *Journal of Cleaner Production*, 101: 271-277.

Chiumenti, A., da Borso, F., Chiumenti, R., Teri, F. and Segantin, P. (2013). Treatment of digestate from a co-digestion biogas plant by means of vacuum evaporation: tests for process optimization and environmental sustainability. *Waste Manage*, 33: 1339-1344.

Dhoble, A. S. and Pullammanappallil, P. C. (2014). Design and operation of an anaerobic digester for waste management and fuel generation during long term lunar mission. *Advances in Space Research*, 54: 1502–1512.

Dias, M. O. S., Junqueira, T. L., Cavalett, O., Cunha, M. P., Jesus, C. D. F., Mantelatto, P. E., Rossell, C. E. V., Maciel Filho, R. and Bonomi, A. (2013). Cogeneration in integrated first and second generation ethanol from sugarcane. *Chemical Engineering Research and Design*, 91: 1411-1417.

Ding, G. K. C. (2017). Wastewater Treatment and Reuse - The Future Source of Water Supply. *Encyclopedia of Sustainable Technologies*, 4: 43-52.

Đurđica, K., Davor, K., Slavko, R., Daria, J., Rober, S. and Marina, T. (2019). Electroporation of harvest residues for enhanced biogas production in anaerobic co-digestion with dairy cow manure. *Bioresource Technology*, 274:215-224.

Dutta, K., Daverey, A. and Lin, J. (2014). Evolution retrospective for alternative fuels: First to fourth generation. *Renewable Energy*, 69: 114-122.

Ek, M., Bergstrom, R., Bjurhem, J.-E., Bjorlenius, B. and Hellstrom, D. (2006). Concentration of nutrients from urine and reject water from anaerobically digested sludge. *Water Science & Technology*, 54: 437-444.

Esteves, E. M. M., Herrera, A. M. N., Esteves, V. P. P. and Morgado, C. R. V. (2019). Life cycle assessment of manure biogas production: A review. *Journal of Cleaner Production*, 219: 411-423.

Fantin, V., Giuliano, A., Manfredi, M., Ottaviano, G., Stefanova, M. and Masoni, P. (2015). Environmental assessment of electricity generation from an Italian anaerobic digestion plant. *Biomass and Bioenergy*, 83: 422-435.

Felca, A. T. A., Barros, R. M., Filho, Dos Santos, I. F. S. and Ribeiro, E. M. (2018). Analysis of biogas produced by the anaerobic digestion of sludge generated at wastewater treatment plants in the South of Minas Gerais, Brazil as a potential energy source. *Sustainable Cities and Society*, 41: 139-153.

Ferella, F., Cucchiella, F., D'Adamo, I. and Gallucci, K. (2019). A techno-economic assessment of biogas upgrading in a developed market. *Journal of Cleaner Production*, 210: 945-957.

Fingerman, K. R., Torn M. S., O'Hare, M. H. and Kammen, D. M. (2010). Accounting for the water impacts of ethanol production. *Environmental Research Letters*, 5: 14-20.

Franchino, M., Comino, E., Bona, F. and Riggio, V. A. (2013). Growth of three microalgae strains and nutrient removal from an agro-zootechnical digestate. *Chemosphere*, 92: 738-744.

Freitas, F. F., De Souza, S. S., Ferreira, L. R. A., Otto, R. B., Alessio, F. J., De Souza, S. N. M., Venturini, O. J. and Ando Junior, O. H. (2019). The Brazilian Market of distributed biogas generation: Overview, technological development and case study. *Renewable and Sustainable Energy Reviews,* 101: 146-157.

Gao, S., Hu, C., Sun, S., Xu, J., Zhao, Y. and Zhang, H. (2018). Performance of piggery wastewater treatment and biogas upgrading by three microalgal cultivation technologies under different initial COD concentration. *Energy*, 165: 360-369.

Ge, S., Madill, M. and Champagne, P. (2018). Use of freshwater macroalgae Spirogyra sp. for the treatment of municipal wastewaters and biomass production for biofuel applications. *Biomass and Bioenergy*, 111: 213-223.

Gerbens-Leenes, W., Hoekstra, A. Y. and Meer, T. H. (2009). The water footprint of bioenergy. *Processing of the National Academy of Science*, 25: 10219-10223.

Gheewala, S. H., Silalertruksa, T., Nilsalab, P., Mungkung, R., Perret, S. R., and Chaiyawannakarn, N. (2013). Implications of the biofuels policy mandate in Thailand on water: The case of bioethanol. *Bioresource Technology*, 150: 457-465.

Gogina, E. and Makisha, N. (2014). Information technologies in view of complex solution of waste water problems. *Applied Mechanics and Materials*, 587-589: 636-639.

Gouveia, L. (2011). Microalgae as a Feedstock for Biofuels. *Springer Briefs in Microbiology*, 68: 1-5.

Greetham, D., Zaky, A., Makanjuola, O. and Du, C. (2018). A brief review on bioethanol production using marine biomass, marine microorganism and seawater. *Current Opinion in Green and Sustainable Chemistry*, 14: 53-59.

Gupta, V. K., Ali, I., Saleh, T. A., Nayak, A., and Agarwal, A. (2012). Chemical treatment technologies for waste-water recycling - an overview. *Royal Society of Chemistry*, 2: 6380-6388.

Hien, P. G., Oanth, L. T. K., Viet, N. T., and Lettinga, G. (1999). Closed wastewater system in the tapioca industry in Vietnam. *Water Science & Technology*, 39: 89-96.

Hjorth, M., Christensen, K. V., Christensen, M. L. and Sommer, S. G. (2010). Solid-liquid separation of animal slurry in theory and practice. *Sustainable Agriculture*, 2: 953-986.

Hoekstra, A. Y., Chapagain, A. K., Aldaya, M. M. and Mekonnen, M. M. (2011). *The water footprint assessment manual: Setting the global standard*. Routledge.

Intanoo, P., Suttikul, T., Leethochawalit, M., Gulari, E. and Chavadej, S. (2014). Hydrogen production from alcohol wastewater with added fermentation residue by an anaerobic sequencing batch reactor (ASBR) under thermophilic operation. *International Journal of Hydrogen Energy*, 39: 9611-9620.

Islam, K. M. N. (2017). Greenhouse gas footprint and the carbon flow associated with different solid waste management strategy for urban metabolism in Bangladesh. *Science of the Total Environment*, 580: 755-769.

Jambo, S. A., Abdulla, R., Azhar, S. H. M., Marbawi, H., Gansau, J. A. and Ravindra, P. (2016). A review on third generation bioetanol feedstock. *Renewable and Sustainable Energy Reviews*, 65: 756-776.

Jang, S., Shirai, Y., Uchida, M. and Wakisaka, M. (2012). Production of mono sugar from acid hydrolysis of seaweed. *African Journal of Biotechnology*, 11: 1953-1963.

Jehlee, A., Khongkliang, P. and O-Thong, S. (2017). Biogas Production from *Chlorella* sp. TISTR 8411 Biomass Cultivated on Biogas Effluent of Seafood Processing Wastewater. *Energy Procedia*, 138: 853-857.

Jung, K. A., Lim, S., Kim, Y. and Park, J. M. (2013). Potentials of macroalgae as feedstocks for biorefinery. *Bioresource Technology*, 135: 1182-190.

Junzhang, L., Bin, H., Gongzhe, C., Jing, W., Yun, F., Xiaoming, T. and Weidong, W. (2014). A study on the microbial community structure in oil reservoirs developed by water flooding. *Journal of Petroleum Science and Engineering*, 122: 354-359.

Kadir, W. N. A., Lam, M. K., Uemura, Y., Lim, J. W. and Lee, K. T. (2018). Harvesting and pre-treatment of microalgae cultivated in wastewater for biodiesel production: A review. *Energy Conversion and Management*, 171: 1416-1429.

Kaewkannetra, P., Imai, T., Garcia-Garcia, F. J. and Chiu, T. Y. (2009). Cyanide removal from cassava mill wastewater using *Azotobactor vinelandii* TISTR 1094 with mixed microorganisms in activated sludge treatment system. *Journal of Hazardous Materials*, 172: 224-228.

Kalakodimi, R. P. and Esmacher, M. J. (2009). Boiler Chemistry Management Using Coordinated Approach of Chemicals, Membranes and Online Monitoring. *GE Water & Process Technologies*, 1-16.

Karuppiah, R., Peschel, A., Grossman I. E., Martín, M., Martinson, W. and Zullo, L. (2008). Energy optimization for the design of corn-based ethanol plants. *AIChE Journal*, 54: 1499-1525.

Khambaty, Y., Upadhyay, D., Kripani, Y., Joshi, N., Mody, K. and Gandhi, M. R. (2013). Bioethanol from macroalgal biomass: utilization of marine yeast for production of the same. *Bioenergy Research*, 6: 188-195.

Khayum, N., Anbarasu, S. and Murugan, S. (2018). Biogas potential from spent tea waste: A laboratory scale investigation of co-digestion with cow manure. *Energy*, 165: 760-768.

Kim, J., Kim, B., Lee, C., Kim, S., Jee, H., Koh, J., and Fane, A. G. (1997). Development of clean technology in alcohol fermentation industry. *Journal of Cleaner Production*, 5: 263-267.

Koszel, M. and Lorencowicz, E. (2015). Agricultural use of biogas digestate as a replacement fertilizers. *Agriculture and Agricultural Science Procedia*, 7: 119-124.

Kouřimská, L., Poustková, I. and Babička, L. (2012). The use of digestate as a replacement of mineral fertilizers for vegetables growing. *Scientia Agriculturae Bohemica*, 43: 121-126.

Kulakov, A. A. (2013). Ecological assessment of the complex water object-treated wastewaters output. *Water Supply and Sanitary Techniques*, 5: 25-30.

Kulakov, A. A. and Lebedeva, E. A. (2011). Development of engineering solutions concerning modernization of wastewater treatment facilities based on technology simulation. *Water Treatment*, 12: 10-19.

Kumar, V., Wati, L., Nigam, P., Banat, I. M., Yadav, B. S., Singh, D., and Marchant, R. (1997). Decolorization and biodegradation of anaerobically digested sugarcane molasses spent wash effluent from biomethanation plants by white-rot fungi. *Process Biochemistry*, 33: 83-88.

Lantz, M., Svensson, M., Björnsson, L. and Börjesson, P. (2007). The prospects for an expansion of biogas systems in Sweden-incentives, barriers and potentials. *Energy Policy*, 35: 1830-1843.

Larsson, M., Galbe, M., and Zacchi, G. (1997). Recirculation of process water in the production of ethanol from softwood. *Bioresource Technology*, 60: 143-151.

Lazaro, C. Z., Varesche, M. B. A. and Silva, E. L. (2015). Sequential fermentative and phototrophic system for hydrogen production: An approach for Brazilian alcohol distillery wastewater. *International Journal of Hydrogen Energy*, 40: 9642-9655.

Ledda, C., Schievano, A., Salati, S. and Adani, F. (2013). Nitrogen and water recovery from animal slurries by a new integrated ultrafiltration, reverse osmosis and cold stripping process: A case study. *Water Research*, 47: 6165-6166.

Lee, P., Bae, J., Kim, J., and Chen, W. (2011). Mesophilic anaerobic digestion of corn thin stillage: a technical and energetic assessment of the corn-to-ethanol industry integrated with anaerobic digestion. *Journal of Chemical Technology and Biotechnology*, 86: 1514-1520.

Leonzio, G. (2016). Upgrading of biogas to bio-methane with chemical absorption process: simulation and environmental impact. *Journal of Cleaner Production*, 131: 364-375.

Lin, C. S. K., Luque, R., Clark, J. H., Webb, C. and Du, C. (2011). A seawater-based biorefining strategy for fermentative production and chemical transformations of succinic acid. *Energy and Environmental Science*, 4: 1471-1479.

Liu, C., Li, K., Wen, Y., Geng, B., Liu, Q. and Lin, Y. (2019). Bioethanol: New opportunities for ancient product. *Advances in Bioenergy,* 01-34.

Liu, J., Liu, Q., and Yang, H. (2016). Assessing water scarcity by simultaneously considering environmental flow requirements, water quantity, and water quality. *Ecological Indicators*, 60: 434-441.

Lu, H., Zhang, G., Zheng, Z., Meng, F., Du, T. and He, S. (2019). Bio-conversion of photosynthetic bacteria from non-toxic wastewater to realize wastewater treatment and bioresource recovery: a review. *Bioresource Technology*, 278: 383-399.

Makisha, N. (2016). Waste water and biogas – Ecology and economy. *Procedia Engineering*, 165: 1092-1097.

Manninen, K., Koskela, S., Antikainen, R., Bocken, N., Dahlbo, H. and Aminoff, A. (2018). Do circular economy business models capture intended environmental value propositions? *Journal of Cleaner Production*, 171: 413-422.

Martín, M., Ahmetovic, E. and Grossmann, I. E. (2011). Optimization of Water Consumption in Second Generation Bioethanol Plants. *Industrial & Engineering Chemistry Research,* 50: 3705-3721.

Mathioudakis, V., Gerbens-Leenes, P. W., Van der Meer, T. H. and Hoekstra, A. Y. (2017). The water footprint of second-generation bioenergy: A comparison of biomass feedstocks and conversion techniques. *Journal of Cleaner Production*, 148: 571-582.

Matsumura, E. M., and Mierzwa, J. C. (2008). Water conservation and reuse in poultry processing plant - A case study. *Resources, Conservation and Recycling*, 52: 835-842.

Meyer, A. K. P., Ehimen, E. A. and Holm-Nielsen, J. B. (2017). Future European biogas: Animal manure, straw and grass potentials for a sustainable European biogas production. *Biomass and Bioenergy*, 111: 154-164.

Mohapatra, S., Ray, R. C. and Ramachandran, S. (2019). "Bioethanol from Biorenewable Feedstocks: Technology, Economics, and Challenges." In: *Bioethanol Production from Food Crops*, edited by Ramesh Ray and S Ramachandran, 3-27. Bhubaneswar: Academic Press.

Moreda, I. L. (2016). The potential of biogas production in Uruguay. *Renewable and Sustainable Energy Reviews*, 54: 1580-1591.

Mosqueira-Salazar, K. J., Palacios-Bereche, R., Chávez-Rodriguez, M., Seabra, J. and Nebra, S. A. (2013). Reduction of water consumption in integrated first-and-second-generation etanol plant. *Energy for Sustainable Development,* 17: 531-535.

Nataraj, S. K., Hosamani, K. M., and Aminabhavi, T. M. (2006). Distillery wastewater treatment by the membrane-based nanofiltration and reverse osmosis processes. *Water Research*, 40: 2349-2356.

Ngumah, C., Ogbulie, J., Orji, J. and Amadi, E. (2013). Potential of organic waste for biogas and biofertilizer production in Nigeria. *Environmental Research, Engineering and Management,* 63: 60–66.

Nguyen, M.-L. T., Lin, C.-Y. and Lay, C.-H. (2019). Microalgae cultivation using biogas and digestate carbon sources. *Biomass and Bioenergy*, 122: 426-432.

Nkoa, R. (2014). Agricultural benefits and environmental risks of soil fertilization with anaerobic digestates: a review. *Agronomy for Sustainable Development*, 34: 473–492.

Oker, T. E., Kisekka, I., Sheshukov, A. Y., Aguilar, J. and Rogers, D. H. (2018). Evaluation of maize production under mobile drip irrigation. *Agricultural Water Management*, 210: 11-21.

Olguín, E. J. (2003). Phycoremediation: key issues for cost-effective nutrient removal processes. *Biotechnology Advances*, 22: 81-91.

Oliveira, C. M., Cruz, A. J. G. and Costa, C. B. B. (2016). Improving second generation bioethanol production in sugarcane biorefineries through energy integration. *Applied Thermal Engineering*, 109: 819-827.

Pina, E. A., Palacios-Bereche, R., Chavez-Rodriguez, M. F., Ensinas, A. V., Modesto, M. and Nebra, S. A. (2017). Reduction of process steam demand and water-usage through heat integration in sugar and ethanol production from sugarcane – Evaluation of diferents plant configurations. *Energy*, 138: 1263-1280.

Poeschl, M., Ward, S. and Owende, P. (2010). Prospects for expanded utilization of biogas in Germany. *Renew and Sustainable Energy Reviews*,14: 1782-1797.

Puyol, D., Batstone, D. J., Hülsen, T., Astals, S. and Krömer, J. O. (2017). Resource recovery from wastewater by biological technologies: opportunities, challenges, and prospects. *Frontiers in Microbiology*, 7: 1-23.

Quan, X., Tao, K., Mei, Y., and Jiang, X. (2014). Power generation from cassava alcohol wastewater: effects of pretreatment and anode aeration. *Bioprocess and Biosystems Engineering*, 37: 2325-2332.

Ramchandran, D., Rajagopalan, N., Strathmann, T. J. and Singh, V. (2013). Use of treated effluent water in ethanol production from cellulose. *Biomass and Bioenergy*, 56: 22-28.

Rawat, I., Kumar, R. R., Mutanda, T., and Bux, F. (2011). Dual role of microalgae: Phycoremediation of domestic wastewater and biomass production for sustainable biofuels production. *Applied Energy*, 88: 3411-3424.

Rawat, S. (2015). Food Spoilage: Microorganisms and their prevention. *Pelagia Research Library*, 5: 47-56.

Rehl, T. and Müller, J. (2011). Life cycle assessment of biogas digestate processing technologies. *Resources, Conservation and Recycling*, 56: 92-104.

Rodriguez, L., Villaseñor, J. and Fernandez, F. J. (2013). Influence of the cleaning additives on the methane production from brewery effluents. *Chemical Engineering Journal*, 215-216: 685-690.

Saravanakumar, K., Senthilraja, P. and Kathiresan, K. (2013). Bioethanol production by mangrove-derived marine yeast Saccharomyces cerevisiae. *Journal of King Saud University*, 25: 121-127.

Scherbakow, V., Gogina, E., Schukina, T., Kuznetsona, N., Makisa, N. and Poupyrey, E. (2015). Calculation of biogas facilities for recycling of organic sewage sludge of breeding factories. *International Journal of Applied Engineering Research,* 10: 44353-44356.

Searmsirimongkol, P., Rangsunvigit, P., Leethochawalit, M. and Chavadej, S. (2011). Hydrogen production from alcohol distillery wastewater containing high potassium and sulfate using an anaerobic sequencing batch reactor. *International Journal of Hydrogen Energy*, 36: 12810-12821.

Shojaosadati, S. A. (1996). The Use of Biomass and Stillage Recycle in Conventional Ethanol Fermentation. *Journal of Chemical Technology and Biotechnology*, 67: 362-366.

Sibilo, S., Rosato, A., Ciampi, G., Scorpio, M. and Akisawa, A. (2017). Building-integrated trigeneration system: energy, environmental and economic dynamic performance assessment for Italian residential applications. *Renewable and Sustainable Energy Reviews*, 68: 920-933.

Silva, JR., Junior, O. H., Spacek, A. D., Mota, J. M., Malfatti, C. F. and Furtado, A. C. (2016). Scaling a biodigester ascendant flow for biogas production via sewer and solid waste. *In: V Proceedings of the International Conference Renew Energies Power Qual*, Madrid (Spain), 517–520.

Singh, J. and Gu, S. (2010). Commercialization potential of microalgae for biofuels production. *Renewable and Sustainable Energy Reviews*, 14: 2596-2610.

Singh, N., Petrinic, I., Hélix-Nielsen, C., Basu, S. and Balakrishnan, M. (2018). Concentrating molasses distillery wastewater using biomimetic forward osmosis (FO) membranes. *Water Research*, 130: 271-280.

Siriwong, T., Laimeheriwa, B., Aini, U. N., Cahyanto, M. N., Reungsang, A. and Salakkam, A. (2019). Cold hydrolysis of cassava pulp and its use in simultaneous saccharification and fermentation (SSF) process for ethanol fermentation. *Journal of Biotechnology*, 292: 57-63.

Sonune, A., and Ghate, R. (2004). Developments in wastewater treatment methods. *Desalination*, 167: 55-63.

Srinivasan, S. (2008). Positive externalities of domestic biogas initiatives: implications for financing. *Renew and Sustainable Energy Reviews*, 12:1476-1484.

Świątczak, P., Cydzik-Kwiatkowska, A. and Zielińska, M. (2019). Treatment of the liquid phase of digestate from a biogas plant for water reuse. *Bioresource Technology*, 276: 226-235.

Tampio, E., Marttinen, S. and Rintala, J. (2016). Liquid fertilizer products from anaerobic digestion of food waste: mass, nutrient and energy balance of four digestate liquid treatment systems. *Journal of Cleaner Production*, 125: 22-32.

Ten Hoeve, M., Hutchings, N. J., Peters, G. M., Svanstrom, M., Jensen, L. S. and Bruun, S. (2014). Life cycle assessment of pig slurry treatment technologies for nutrient redistribution in Denmark. *Journal of Environmental Management*, 132: 60-70.

Törnwall, E., Pettersson, H., Thorin, E. and Schwede, S. (2017). Post-treatment of biogas digestate - An evaluation of ammonium recovery, energy use and sanitation. *Energy Procedia* 142: 957–963.

Tyagi, S., Lee, K., Mulla, S. I., Garg, N. and Chae, J. (2019). Production of Bioethanol From Sugarcane Bagasse: Current Approaches and

Perspectives. *In:* Applied Microbiology and Bioengineering, 1. ed, 278p.

Uusitalo, V., Soukka, R., Horttanainen, M., Niskanen, A. and Havukainen, J. (2013). Economics and greenhouse gas balance of biogas use systems in the Finnish transportation sector. *Renewable* Energy, 51:132-140.

Vaccari, G. Tamburini, E., Sgualdino, G., Urbaniec, J. and Klemes, J. (2005). Overview of the environmental problems in beet sugar processing: possible solutions. *Journal of cleaner production*, 13: 499-507.

Van Stappen, F., Mathot, M., Decruyenaere, V., Loriers, A., Delcour, A., Planchon, V., Goffart, J. and Stilmant, D. (2016). Consequential environmental life cycle assessment of a farm-scale biogas plant. *Journal of Environmental Management*, 175: 20-32.

Vanham, D., Hoekstra, A. Y., Wada, Y., Bouraoui, F., Roo, A., Mekonnen, M. M., Bund, W. J. V., Batelaan, O., Pavelic, P., Bastiaanssen, W. G. M., Kummu, M., Rockström, J., Liu, J., Bisselink, B., Ronco, P., Pistochi, A. and Bidoglio, G. (2018). Physical water scarcity metrics for monitoring progress towards SDG target 6.4: An evaluation of indicator 6.4.2 "Level of water stress." *Science of the Total Environment*, 613-614: 218-232.

Vergine, P., Salerno, C., Libutti, A., Beneduce, L., Gatta, G., Berardi, G. and Pollice, A. (2017). Closing the water cycle in the agro-industrial sector by reusing treated wastewater for irrigation. *Journal of Cleaner Production*, 164: 587-596.

Vohra, M., Manwar, J., Manmode, R., Padgilwar, S. and Patil, S. (2014). Bioethanol production: Feedstock and current technologies. *Journal of Environmental Chemical Engineering*, 2: 573-584.

Volkov, A., Chulkov, V., Kazaryan, R., Fachratoy, M., Kyzina, O. and Gazaryan, R. (2014). Components and guidance for constructional rearrangement of buildings and structures within reorganization cycles. *Applied Mechanics and Materials*, 580-583: 2281-2284.

Wan, Y. K., Sadhukhan, J., and Ng, D. K. S. (2015). Techno-Economic Evaluation of Integrated Bioethanol Plant for Sago Starch Processing Facility. *Chemical Engineering Research and Design*, 107: 102-116.

Wang, M., Kuo-Dahab, W. C., Dolan, S. and Park, C. (2014). Kinetics of nutrient removal and expression of extracellular polymeric substances of the microalgae, *Chlorella* sp. and *Micractinium* sp., in wastewater treatment. *Bioresource Technology*, 154: 131-137.

Wang, X., Gao, S., Zhang, Y., Zhao, Y. and Cao, W. (2017). Performance of different microalgae-based technologies in biogas slurry nutrient removal and biogas upgrading in response to various initial CO2 concentration and mixed light-emitting diode light wavelength treatments. *Journal of Cleaner Production*, 166: 408-416.

Wang, Z., Zhao, Y., Ge, Z., Zhang, H. and Sun, S. (2015). Selection of microalgae for simultaneous biogas upgrading and biogas slurry nutrient reduction under various photoperiods. *Journal of Chemical Technology & Biotechnology*, 91: 1982-1989.

Wu, M. and Chiu, Y. (2011). Consumptive water use in the production of ethanol and petroleum gasoline. USA: Transportation Technology R&D Center.

Wu, M., Mintz, M., Wang, M. and Arora S. (2009). Water consumption in the production of ethanol and petroleum gasoline. *Environmental Management*, 44: 981-997.

Wu, M., Mintz, M., Wang, M. and Arora, S. (2009). Consumptive Water Use in Major Steps of the Cellulosic Ethanol Lifecycle. In *Consumptive Water Use in the Production of Ethanol and Petroleum Gasoline,* 28-29.

Xia, A. and Murphy, J. D. (2016). Microalgal Cultivation in Treating Liquid Digestate from Biogas Systems. *Trends in Biotechnology*, 34: 264-275.

Xin, Y., Sun, B., Zhu, X., Yan, Z., Zhao, X., and Sun, X. (2017). Resourceful treatment of alcohol distillery wastewater by pulsed discharge. *Bioresource Technology*, 244: 175-181.

Xu, J., Zhao, Y., Zhao, G. and Zhang, H. (2015). Nutrient removal and biogas upgrading by integrating freshwater algae cultivation with piggery anaerobic digestate liquid treatment. *Applied Microbiology and Biotechnology*, 99: 6493-6501.

Yang, L., Tan, X., Li, D., Chu, H., Zhou, X., Zhang, Y. and Yu, H. (2015). Nutrients removal and lipids production by *Chlorella pyrenoidosa* cultivation using anaerobic digested starch wastewater and alcohol wastewater. *Bioresource Technology*, 181: 54-61.

Yang, Q. and Chen, G. Q. (2013). Greenhouse gas emissions of corn–ethanol production in China. *Ecological Modelling*, 252: 176-184.

Yang, X., Wang, K., Wang, H., Zhang, J. and Mao, Z. (2016). Ethanol fermentation characteristics of recycled water by *Saccharomyces cerevisiae* in an integrated ethanol-methane fermentation process. *Bioresource Technology*, 220: 609-614.

Yavuz, Y. (2007). EC and EF processes for the treatment of alcohol distillery wastewater. *Separation and Purification Technology*, 53: 135-140.

Zaky, A. S., Greetham, D., Louis, E. J., Tucker, G. A. and Du, C. (2016). A new isolation and evaluation method for marine-derived yeast spp. with potential applications in industrial biotechnology. *Journal of Microbiology and Biotechnology*, 26: 1891-1907.

Zaky, A. S., Tucker, G. A., Daw, Z. Y. and Du, C. (2014). Marine yeast isolation and industrial application. *FEMS Yeast Research*, 14: 813-825.

Zeng, Y., Xiao, L., Zhang, X., Zhou, J., Ji, G., Schroeder, S., Liu, G. and Yan, Z. (2018). Biogas desulfurization under anoxic conditions using synthetic wastewater and biogas slurry. *International Biodeterioration & Biodegradation*, 133: 247-255.

Zhang, Q., Lu, X., Tang, L., Mao, Z., Zhang, J., Zhang, H. and Sun, F. (2010). A novel full recycling process through two-stage anaerobic treatment of distillery wastewater for bioethanol production from cassava. *Journal of Hazardous Materials*, 179: 635-641.

Zhang, T., Chi, Z., Zhao, C. H., Chi, Z. M. and Gong, F. (2010). Bioethanol production from hydrolysates of inulin and the tuber meal of Jerusalem artichoke by Saccharomyces sp. W0. *Bioresource Technology*, 101: 8166-8170.

Zhao, C., Chen, B. and Yang, J. (2014). Embodied water consumption of biogas-digestate utilization. *Energy Procedia*, 61: 615-618.

Zhu, L. (2015). Microalgal culture strategies for biofuel production: a review. *Biofuels, Bioproducts and Biorefining*, 9: 801–814.

Zuliani, L., Frison, N., Jelic, A., Fatone, F., Bolzonella, D. and Ballottari, M. (2016). Microalgae Cultivation on Anaerobic Digestate of Municipal Wastewater, Sewage Sludge and Agro-Waste. *International Journal of Molecular Sciences*, 17: 1692.

In: Wastewater Treatment
Editor: Adriana Magdalena

ISBN: 978-1-53616-370-4

Chapter 5

PETROLEUM WASTEWATER TREATMENT USING GRANULAR SEQUENCING BATCH REACTOR: PHYSICAL CHARACTERISTICS AND CAPABILITIES OF THE AEROBIC GRANULES

Shabnam Taghipour and Bita Ayati*, PhD
Civil and Environmental Engineering,
Tarbiat Modares University, Tehran, Iran

ABSTRACT

In general, petroleum wastewater has high concentration of organic and inorganic components such as BTEX, MTBE, PAH, phenol, ammonia, sulfides, cyanides, and heavy metals which make it necessary to be treated before releasing to the environment. Granular sequencing batch reactor is one of the promising biological methods for cultivating aerobic granules and consequently treating wastewater. Owing to numerous advantages (i.e., stronger microbial structure of the granules,

* Corresponding Author's E-mail: ayati_bi@modares.ac.ir.

high removal efficiency, sludge dewatering ability, appropriate settle-ability, and high biomass maintenance and resistance against organic loading shocks), this technology has been used in remediation of various pollutants from aqueous solutions. In this chapter, characteristics of the cultivated bio-granules including number of granules, density, sludge volume index, settling velocity, and physical strength have been studied and their performance in treating petroleum and other types of wastewater have been investigated.

Keywords: aerobic granules, petroleum, wastewater, organic loading, granular sequencing batch reactor (GSBR)

1. Introduction

Annually, 1012 US gallons of petroleum hydrocarbons are consumed for production of more than 2500 petroleum products in all over the world [1]. Based on composition of crude oil, type and complexity of operation and the products, petroleum refineries can produce a large amount of wastewater with toxic compounds [2]. Petroleum wastewater (PWW) generally contains organic compounds such as oil and grease, mercaptans, heavy oil (viscosity > 100 mPas), polycyclic aromatic hydrocarbons (PAH), phenols, and inorganic compounds such as sulfides, chlorides, ammonia (NH3) and heavy metals [3-14]. PWW pollution can affect different aspects of human life such as [15]: (1) infecting drinking water and aquatic resources; (2) threatening human health; (3) polluting the atmosphere; (4) affecting crop production; (5) affecting the natural landscape, and (6) enhancing fire triggering probability. Therefore; finding a cost-effective treatment method has attracted direct attention by the researchers. Different treatment methods which have been applied for treatment of oily wastewaters are shown in Figure 1. These methods are: conventional treatment methods of PWW including different physical methods such as adsorption, physical separation, membrane filtration, flotation [16], centrifuge, freezing [17], chemical methods such as coagulation and flocculation [18], electrochemical processes, advanced oxidation processes (AOPs) [19], and biological methods such as activated sludge, up-flow anaerobic sludge blanket (UASB), sequencing batch reactor (SBR), and moving bed biofilm reactor (MBBR) [20]. Some of these technologies consume high energy, need large operational costs, have low treatment efficiency, and generate secondary pollution (i.e., toxic by-products). Economical and eco-friendliness advantages of biological treatments, place them among the most favorable methods in PWW treatment.

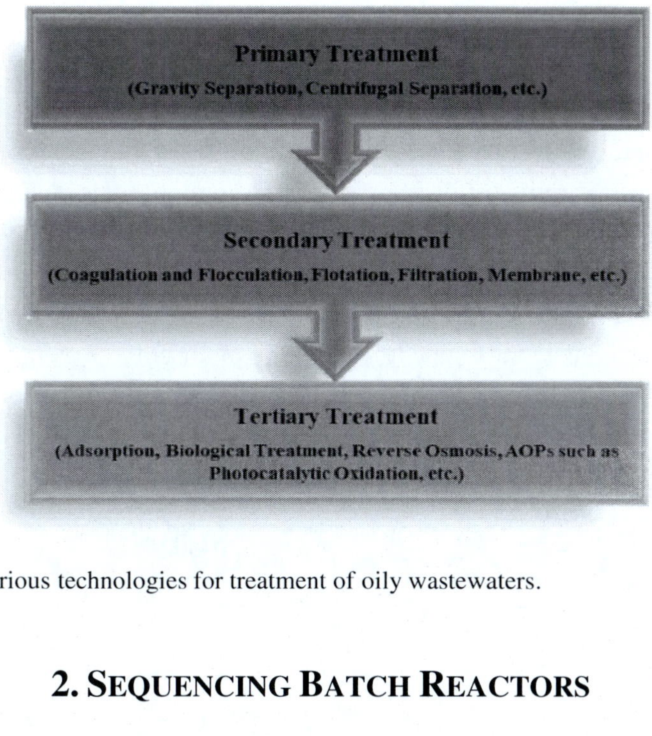

Figure 1. Various technologies for treatment of oily wastewaters.

2. Sequencing Batch Reactors

Conventional activated sludge (CAS) is well-known as a standard technology in biological treatment of various wastewaters. Suspended flocs which are originated from microbial community are the main agents of wastewater treatment in this technology [22]. This technology suffers from low settling-ability (i.e., sludge bulking) which leads to decrease in effluent quality by exceeding loss of sludge in the effluent, lowering treatment efficiency, and also having uncontrollable sludge ages [23]. Developing new technology based on cultivating compact and dense granules compensate some of the CAS drawbacks. Aerobic granular sludge (AGS) was first cultivated in a bubble-column lab-scale reactor in 1997 and successfully applied in treatment of domestic, industrial and high organic loading rate (OLR) wastewaters, bioremediation (biotransformation) of toxic aromatic compounds, adsorption of heavy metals, and so on [24]. Aerobic granules can be cultivated in SBR which is a modified activated sludge process. Therefore; this technology is also known as granular sequencing batch reactor (GSBR) which benefits from advantages such as regular, stronger microbial and compact hydraulically structure of the granules, high nutrient removal efficiency, enhanced sludge dewatering ability, high settle-ability, high biomass maintenance, and high tolerance against changes in pollution loads and shocks [25].

There are various effective factors in granulation including: seed sludge, substrate composition, organic loading rate, reactor structure, duration and rate of settlement, hydrodynamic shear force, cycle duration, feast-famine regime, environmental factors (i.e., pH and temperature), volumetric exchange rate, and the amount of dissolved oxygen [26-30]. GSBR is a cyclic process in which each cycle comprises five phases: fill, react, settle, draw, and idle as shown in Figure 2 [31].

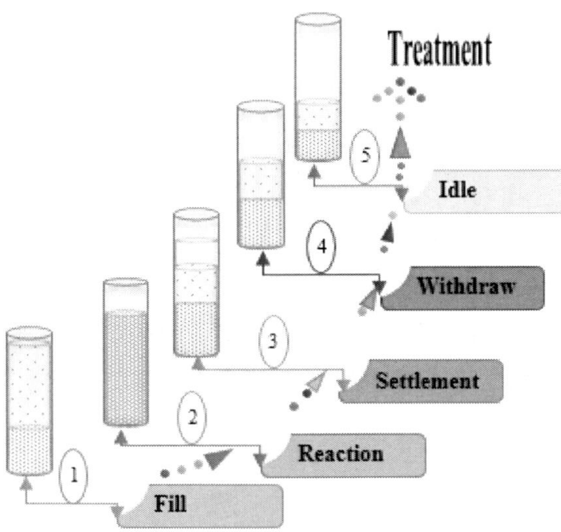

Figure 2. Typical stages in each cycle of a GSBR system.

Successful application of GSBR technology has been reported in several studies. Nancharaiah et al. (2018) investigated treatment of ammonium and selenium rich wastewaters with granular SBR. The system was run on a 24-h cycle including: filling (10 min), anaerobic (8 h), aeration (15 h), settling (5 min), drawing (10 min) and idle (35 min). Lactate (3.74 mM/L.d), NH_4Cl (1.67 mM/L.d), and sodium selenite (10-100 mM/L.d) were used as carbon, ammonium, and selenium sources, respectively. During 200 days of operation the total COD, ammonium, total nitrogen and Se removal efficiencies were in the range of 88-100%, 92-100%, 81.4-97.7%, and 82-98.5%, respectively. They reported the following mechanism for selenite removal by AGS: 1) bio-sorption of Se (IV), 2) bio-reduction of Se (IV) to Se (0) by microorganisms, 3) formation of Se (0) nano-spheres inside the granules, and 4) propagation of Se (0) or Se (0)-containing cells/cell aggregates in the aqueous solution [32].

Hamza et al. reported successful treatment of high-strength organic wastewater (OLR > 30 kg/m^3d) with granular SBR. The composition of the synthetic wastewater was as follows: NaAc anhydrous (2930 mg/L), NH_4Cl (350 mg/L), K_2HPO_4 (30 mg/L), KH_2PO_4 (25 mg/L). The reactor was operated for 100 days in 4-h cycles including: filling (8 min), aeration, settling (20-8 min), and drawing (2 min). Initial COD in first 40 days and from 41 until the end of operation time was equal to 2600 ± 450 mg/L (10.2 ± 2.1 kg COD/m^3.d) and 7500 ± 600 mg/L (27.0 ± 3.5 kg COD/m^3.d), respectively. The maximum removal efficiency in these two stages was 98.4 ± 1.1% and 96 ± 2.7%, respectively. After completing granulation, initial sludge volume index (SVI) (155.5 mL/g) was decreased to below 50 mL/g [33].

Besides several pollutants, application of this technology has been also investigated in treatment of PWW.

3. Chemical Composition of Petroleum Wastewater

Petroleum wastewaters contain wide range of organic and inorganic pollutants. These pollutants occur in different concentrations with different toxicity levels which should be treated before discharging the environment. Although the chemical composition of petroleum varies in different depth, location and age of the field [33], but generally comprises of heavy oil, bitumen, nutrients (nitrogen and phosphorus compounds), metals, oxygen-containing and sulfur-containing compounds [4]. From molecular point of view, molecular compositions of petroleum are categorized in three main groups including: I) aromatics, II) saturates (cycloalkanes, normal and branched alkanes), and III) heteroatoms-containing compounds (nitrogen, sulfur, and oxygen). Approximately 0.1–2 weight percent of crude oils are nitrogen-containing compounds. Nitrogen and oxygen-containing compounds are mainly in polar forms (i.e., phenols, pyridines, carboxylic acids and amides) as compared to nonpolar forms (i.e., ethers) [34]. Oxygen-containing compounds are naphthenic acids (or their salts), phenols (or phenolic compounds), fatty acids, and inclusions in asphaltenes. Metals are other components of crude oil which can be found in oil in the following forms: organic metals, inorganic salts, metal soaps, and attached to asphaltenes [35]. Sulfur-containing compounds in crude oil consist of mercaptans or thiols, sulfides, and thiophenes [26]. Hydrocarbon components of petroleum can be classified into three main groups of [34]:

1. paraffinics (C_nH_{2n+2}, saturated hydrocarbons with chains and without ring structure)
2. naphthenics or alicyclic hydrocarbons (C_nH_{2n}, saturated hydrocarbons with one or more rings and one or more paraffinic side chains)
3. aromatics (consist of six carbon atoms ring, one or more benzene, naphthalene, and phenanthrene ring systems)

4. Petroleum Contaminants

Petroleum wastewater, generally contains 300–600 mg/L of chemical oxygen demand (COD), 150-250 mg/L of biological oxygen demand (BOD), 20–200 mg/L of phenol, 1–100 mg/L of benzene, 1–100 mg/L of benzo (a) pyrene, heavy metals including: 0.1–100 mg/L of chrome, 0.2–10 mg/L of lead, nickel, mercury [36] and other pollutants [21, 37-40]. Standard limitation values of these parameters and compounds for discharging the environment are as follow: 125 mg/L for COD, 15 mg/L for BOD, 0.35 mg/L for phenol, 0.1 mg/L for benzene, 0.2 mg/L for benzo (a) pyrene, 2.0 mg/L for total chromium, 0.1 mg/L for lead, 1 mg/L for nickel, and 0.01 mg/L for mercury [21, 41].

All of the above mentioned compounds either organic or inorganic contaminants can be found in PWW and need to be removed prior to discharge.

4.1. Organic Contaminants

Presence of organic pollutants in aquatic environment reduce the dissolved oxygen for aquatic organisms. Organic oil pollutants such as BTEX (benzene, toluene, ethyl-benzene and xylene) [42-44], MTBE (methyl tetra-butyl ether) [30], PAH (polycyclic aromatic hydrocarbons) [45], and phenol [46] have been successfully removed by GSBR.

Taghipour et al. studied the ability of GSBR in removal of MTBE. They used a GSBR with circular cross-section in 4 h cycle. Considering the changes of granules, optimal inlet COD and removal efficiency were equal to 500 mg/L and 90 percent, respectively. To evaluate the withstanding ability of the bioreactor to organic shock loading rate, the inlet COD was increased to 2000 mg/L in one cycle. After 45 cycles, over 83% removal was observed at the steady state. The first granules were observed in the seventh day of operation and reached the maximum size of 6 mm after 120 days [30].

Chiavola et al. investigated the technical feasibility of GSBR for treatment of PAH contaminated sediments (including: 3-rings (fluorine and anthracine) and 4-rings PAHs (pyrene and crysene)). The final removal efficiency and inlet concentration of PAH were equal to 80 percent and 70 mg/kg (as dry weight), respectively [45].

Yusoff et al. evaluated biodegradation of phenol by hybrid growth sequencing batch reactor. When inlet phenol reduced from 1000 to 200 mg/L, the specific oxygen uptake rate reduction was approximately 20%. At the same time, after 7 h of operation, 93.3% of COD removal efficiency was recorded when the initial COD was equal to 1425 mg/L (C_{phenol} = 600 mg/L). They found that by increasing in the phenol concentration, the toxicity of phenol, leads to decrease in the required oxygen amount for microbial activities [46].

4.2. Inorganic Contaminants

Contaminants such as ammonia (NH_4^+), sulfides, cyanides, and heavy metals are of main inorganic pollutants in different petroleum refining wastewaters [4, 47-50]. Sekine et al. investigated long fill period nitrification of sulfide-containing anaerobic digestion wastewater with GSBR in a long sludge retention time of 139 days. During the first phase (days: 0–18, $C_{sulfide}$ = 0), potential oxidation of ammonium to nitrate was observed on day 6th of experiment. During the second phase (days: 18–30, $C_{sulfide}$ = 32 mg-S/L.d), the nitrification efficiency reduced to 79%, but complete nitrification achieved within two days. Evaluating the sulfide tolerance of the sludge in the GSBR exhibited that the half-maximum inhibitory concentration (IC_{50}) for the sludge has a typical value of 0.73 mg-S/L. They found that long sludge retention time leads to stable nitrification of the microorganisms even under sulfide supply [49].

El-Sheekh et al. evaluated the bioremediation of heavy metals by algal cell immobilization technique on GSBR. The removal efficiency of immobilization technique for total dissolved solids (TDS), COD, BOD, oil and grease, and NH_3 were 51, 97, 98, 98, and 98 percent, respectively. Also removal efficiency of Cr, Ni, Cu, Cd, and Pb was improved by 57, 56, 53, 100, and 100 percent, respectively. At the final stage, the SBR-modified with algal cell immobilization technique (2 h before discharging to non-fresh surface water systems) potentially enhanced the quality of treated wastewater [50].

5. CHARACTERISTICS OF THE GRANULES

Characteristics of the granules such as the number of granules, density, SVI, settling velocity, and physical strength play crucial role in biodegradation of the pollutants by GSBR.

5.1. The Number of Granules

By granules formation over time, the number of granules can be counted. Purification quality of the AGS was enhanced by increase in the number of granules. The number of granules (n) can be calculated via several methods. Some of them are mentioned below:

In this method (Eq. 1), the below-mentioned steps should be followed [51]:

- Harvesting and draining the granular sludge
- Measuring bulk volumes of the total granules in the whole reactor (V, ml)
- Assuming the granule spherical in shape with an average mean radius (r, mm):

$$n = \frac{V \times 1000}{\frac{4}{3}\pi r^3} \tag{1}$$

Using statistical chart (Figure 3). In this method, the below-mentioned steps have been followed [25]:

- Sampling (50 cc) during reaction phase from the sampling valves
- Counting the number of granules in the sample
- Attribute the total volume of the reactor (3.6 L) by statistical diagram (boxplot)

Taghipour and Ayati reported increase in the number of granules on days 26 and 140, from averagely 28 to 96 granules (size > 2mm).

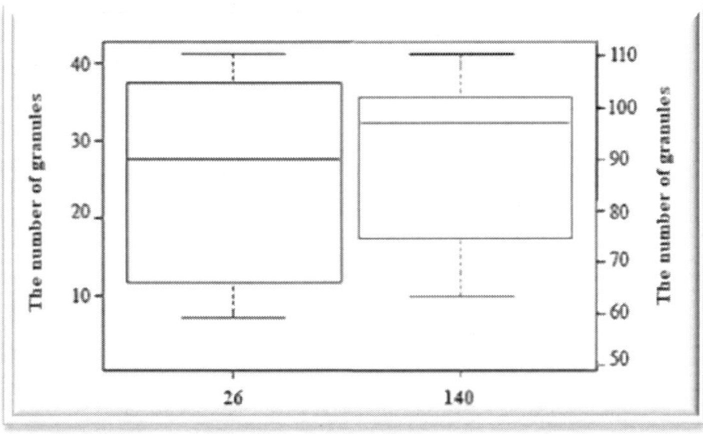

Figure 3. The number of granules in 50cc sample on days 26 and 140. Adapted from ref. 25.

5.2. Morphology of the Granules

Investigating the changes of flocculent samples revealed that the morphology of microbial sludge entirely changed during the experiments. The seed sludge generally has a fluffy and irregular structure with gray color. Over time and by gradually adaptation of microorganisms to the inlet pollutant, the loose structure of the flocs changed to the brown aggregates as shown in Figure 4 [2].

Morphological studies of the granules have been performed largely by transmission electron microscope (TEM) and scanning electron microscopy (SEM) [52-56]. The microbial distribution in a bio-granule generally depends on the decomposition nature of the substrate and type of the carbon source [57]. The aerobic granules cultivated with glucose (as carbon source), are mainly consisted of filamentous and cocci bacteria. While acetate is the carbon source, bacilli bacteria are the dominant bacteria inside the bio-granules [25, 58]. Filamentous-rich granules generally have loose microbial structure. TEM images of the cultivated granules in petroleum wastewater-containing GSBR which was fed by glucose as initial carbon source are presented in Figure 5. Various bacilli, cocci, filamentous, biological masses, and oil-eating bacteria can be seen clearly [2, 25].

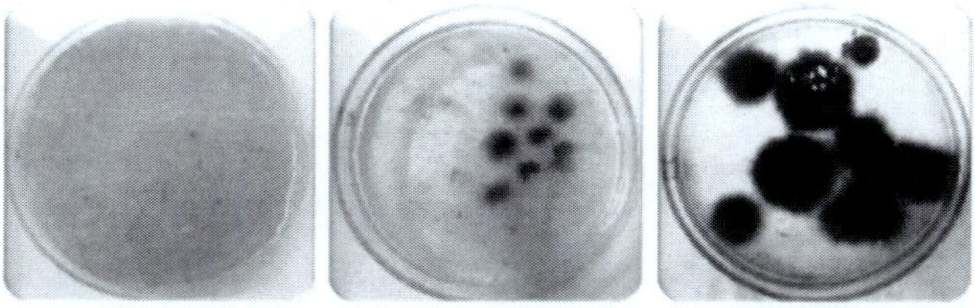

Figure 4. Changes in seed sludge and formation of granules over time. Adapted from ref. 2, with permission from Desalination and Water Treatment. Copyright 2016 Desalination Publications.

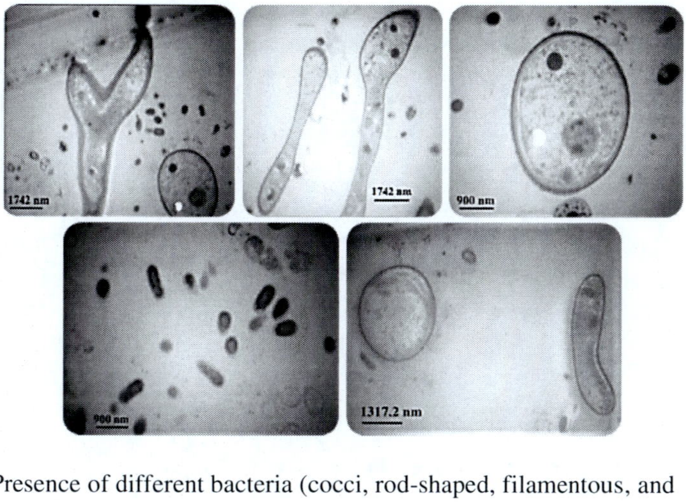

Figure 5. Presence of different bacteria (cocci, rod-shaped, filamentous, and alcanivorax) in TEM images of a mature granule after 150 days of formation. Adapted from ref. 2, with permission from Desalination and Water Treatment. Copyright 2016 Desalination Publications.

5.3. Density of the Granules

Measuring density is a kind of microbial compaction indicator and provides useful information on the settling behavior of granules. In the other words, the granules with high density exhibit strong shear and abrasion resistance. The specific gravity of aerobic granules is usually in the range of 1.004-1.065 [59]. Several factors such as aeration rate, existence of baffles and enough hydrodynamic shear force play crucial role in compaction of the biogranules. If C:N:P ratios (e.g., 100:5:1 on a molar basis) and also the required amount of nutrients, adjusted at the standard range before entering the reactor, the density of the granules will increase over time. While insufficient amount of OLR will lead to decrease in density of aerobic granules. Taghipour and Ayati [2], reported successful cultivation of granules with high density in petroleum wastewater-containing GSBR. They found that OLR more than 2.2 kg/m^3.d, has negatively affected settling behavior of the granules (Figure 6).

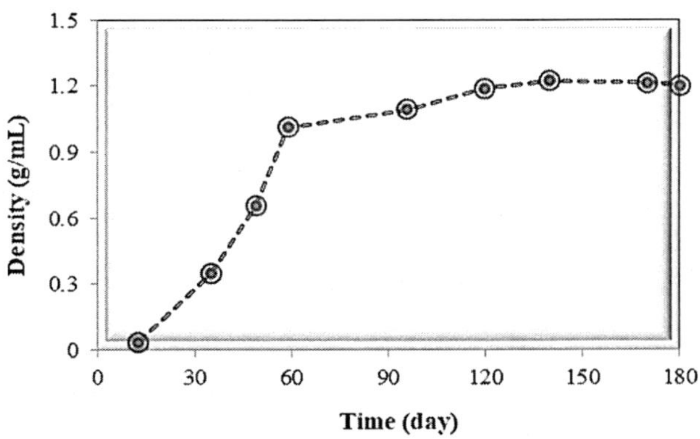

Figure 6. Changes in the density of the cultivated granules in petroleum wastewater over time (OLRin = 0.8–2.4 kg/m^{3}d). Adapted from ref. 2, with permission from Desalination and Water Treatment. Copyright 2016 Desalination Publications.

5.4. Sludge Volume Index (SVI)

Sludge settling ability can be evaluated by sludge volume index (SVI). High amount of SVI (>150 mL/g) is indicator of bulking sludge status which leads to observation of high amount of floated sludge in the effluent

[60]. Owing high settling velocity and low SVI, aerobic granules need no carrier material. Measuring SVI after 5 and 30 minutes is the main criterion for assessing the quality of granulation [61]. SVI30 of mature granules often reported to be in the range of 20–60 ml/g [2]. Kang and Yuan [62] reported decrease in SVI_5 from 111 ± 3 to 55 ± 6 mL/g after 33 days in GSBR system with high organic loads (COD = 400 mg/L). They reported successful granulation (mean diameter = 200μm) with $SVI_5 < 50$ mL/g after 40 days. At the end of day 69th, SVI_5 of bio-granules (diameter >2 mm) were approximately 46 ± 2 mL/g. Cofréa et al. [63] investigated treatment of domestic wastewater (CODin = 550–650 mg/L) in a 4.5 h-cycle GSBR. The sedimentation property of initial seed sludge was characterized by SVI_{30} of 200 mL/g TSS. They recorded 70 mL/g TSS of SVI_{10} for cultivated bio-granules after 30 days operation with mean diameters between 1.0 and 5.0 mm. Taghipour reported potential reduction of SVI after successful granulation in a petroleum wastewater-containing GSBR (Figure 7). SVI was recorded at the end of aeration phase. SVI_{30} and SVI_5 of seed sludge (277 and 297 mL/g) gradually decreased to 38 and 52 mL/g within 90 days, respectively [25].

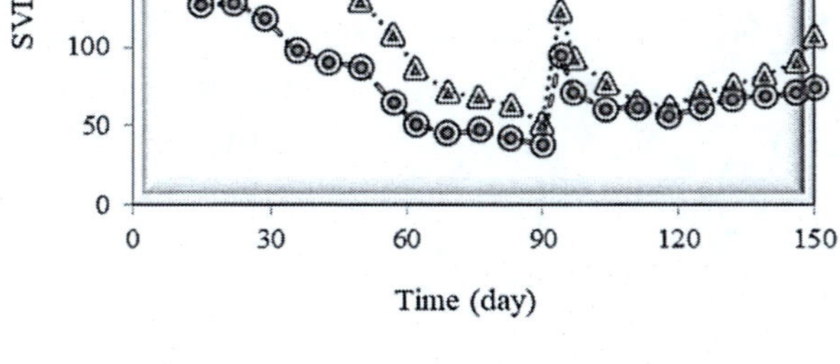

Figure 7. Variation of SVI_{30} and SVI_5 during the operation period. Adapted from ref. 2, with permission from Desalination and Water Treatment. Copyright 2016 Desalination Publications.

5.5. Settling Velocity of the Granules

Settling property of biogranules can be affected by various factors such as properties of fluid and flow, and density and size of the granules. Gravity, buoyancy and drag forces are three main forces involved in granules settlement (Eqs. 2-4) [64]. Stokes flow (Re < 1) usually used for evaluating settling velocity of granules in a constant-velocity free-settling process [65, 66]:

$$\text{Gravity force: } G = \frac{\pi d_m^3 \rho_p g}{6} \tag{2}$$

$$\text{Buoyancy force: } F_b = \frac{\pi d_m^3 \rho_w g}{6} \tag{3}$$

$$\text{Drag force: } F_d = C_d \frac{\pi d_m^2}{4} . \frac{\rho u^2}{2} \tag{4}$$

where d_m, ρ_p, ρ_w, C_d and u are equivalent diameter of sphere with same volume of the granule, density of granule, density of liquid, drag coefficient and terminal settling velocity, respectively. By balancing the forces ($G = F_b + F_d$) and substituting drag coefficient based on Reynolds number, the Stokes equation for determining settling velocity of granules can be obtained as Eq. 5:

$$u = \frac{d_m^2 g(\rho_p - \rho_w)}{18\mu} \tag{5}$$

Another method applied for determining the settling velocity is recording the required time for an individual granule to traverse a certain height in a measuring cylinder filled with tap water [67] as can be seen in Figure 8.

Long et al. operated GSBR system in 6 h per cycle for treatment of wastewater (containing KH_2PO_4, CH_3COONa, NH_4Cl, $CaCl_2$, $FeSO_4.7H_2O$, $MgSO_4$ and several minor elements) with maximum COD equal to 1703.74 mg/L. They used both inoculated activated sludge (75% w/w) and granular sludge (25% w/w) in the reactor. After 4 min of settling time, average setting velocity of granules was approximately 1.22 cm/s. After adaptation of activated sludge, removal efficiency was always above 92% [68]. Dahalan et al. reported maximum settling velocity of 1.11 cm/s for granule (mean diameter = 2.0 mm). They attributed such high settling velocity to regular, compact and smooth morphology of the granules [69]. Taghipour and Ayati reported high settling velocity (2.51 cm/s) for mature granules with average diameter of 6 mm which were cultivated in a petroleum wastewater-containing GSBR. The settling velocity of the granules on days 59, and 86 were 2.61, and 3 cm/s, respectively. They reported even higher values (in the range of 3.12–3.19 cm/s) for the rest of the experiments (days 94 to 180) [2].

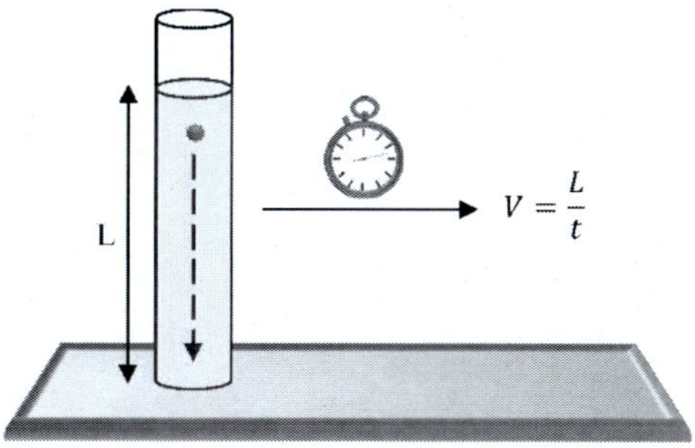

Figure 8. Schematic drawing of measuring settling velocity of cultivated aerobic granules in a GSBR system.

5.6. Physical Strength of the Granules

The physical stability of the granules can be expressed indirectly by integrity coefficient (%IC). This parameter can affect the compactness and activity of the biomass in a reactor. In order to determine the integrity coefficient, the granular sludge should be shaken for 5 min at 200 rpm with an orbital platform shaker. By dividing residual granules to the total weight of granular sludge, the physical strength of the granules can be investigated [67], [70]. The lower values of IC, represent the higher strength of granules. Rosman et al. cultivated aerobic granules in GSBR system for treating rubber wastewater at three different hydraulic retention times

(HRT) of 6, 12, and 24 h. The average sizes, settling velocity, density and IC of the cultivated granules at HRT of 6 h were 2.0 ± 0.1 mm, 1.69 cm/s, 78.2 g/L, and 9.54%. At HRT of 12 h, above-mentioned parameters were 1.2 ± 0.1 mm, 0.92 cm/s, 57.1 g/L, and 15.83%, respectively. By increasing HRT to 24 h, average size, settling velocity, and density of the granules decreased to 0.8 ± 0.1 mm, 0.67 cm/s, 42.3 g/L and IC increased to 18.76%, respectively. The results demonstrated that the lower HRT leads to cultivation of granules with higher physical quality [67]. There is an inverse correlation between IC and density in which increasing in density leads to decrease in IC as can be seen in Figure 9 [2].

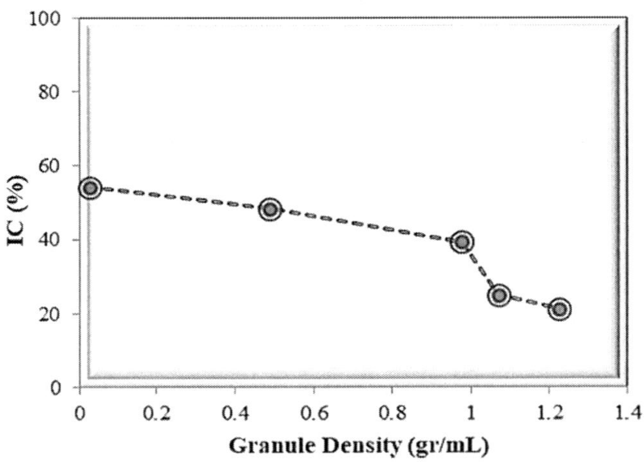

Figure 9. Variation of integrity coefficient (IC) vs. density of the aerobic granules. Granules were cultivated in a petroleum wastewater-containing GSBR. Adapted from ref. 2, with permission from Desalination and Water Treatment. Copyright 2016 Desalination Publications.

Conclusion

Owing various advantages (e.g., compact and stronger microbial structure of the granules, high nutrient removal, high sludge dewatering ability, high settle-ability and biomass maintenance, high tolerance against changes in pollution loads and shocks) aerobic granulation has attracted researchers' attention and become one of the most remarkable technologies in treating wide types of wastewater. Aerobic granules are mainly cultivated in sequencing batch reactors. The microbial distribution in a granule strongly depends on the type of the carbon source. Filamentous and cocci are the dominant microorganisms in cultivated granules with glucose (as carbon source). Aerobic granulation technology has been successfully used for the treatment of high OLR wastewaters (e.g., oily wastewaters). Permanent entrance of oily wastewater into GSBR leads to domination of filamentous in the structure of granules which lead to compactness of the granules. Aeration rate, existence of baffles and enough hydrodynamic shear forces are the other important agents in compactness of the granules. High density and settling velocity of the granules are considered as criteria for successful granulation. Besides above mentioned advantages, this technology suffers from unpredictable disintegration of the granules in high OLR. Moreover, application of this technology in full-scale and finding efficient way for rapid cultivation of granules are another important on-going challenges of the researches for this technology. Therefore, further studies are necessary to find out an approach for these issues.

REFERENCES

[1] El-Naas, Muftah H., Manal Abu Alhaija, and Sulaiman Al-Zuhair. 2014. "Evaluation of a three-step process for the treatment of petroleum refinery wastewater." *Journal of Environmental Chemical Engineering*, 2(1):56-62. doi: 10.1016/j.jece.2013.11.024.

[2] Taghipour, Shabnam, and Ayati, Bita. 2017. "Cultivation of aerobic granules through synthetic petroleum wastewater treatment in a cyclic aerobic granular reactor." *Journal of Desalination and Water Treatment*, 76:134-142. doi: 10.5004/dwt.2017.20779.

[3] Ani, Ijeoma J., Akpan, Uduak G., Olutoye, MOSES A. and Hameed, Bassim H. 2018. "Photocatalytic degradation of pollutants in petroleum refinery wastewater by TiO_2 and ZnO-based

photocatalysts: Recent development." *Journal of Cleaner Production*, 205:930-954. doi: 10.1016/j.jclepro.2018.08.189.

[4] Mustapha, Hassana I., and Piet N.L. Lens. 2018. "Constructed Wetlands to Treat Petroleum Wastewater." *In Approaches in Bioremediation*. Springer, Cham.

[5] Mallick, S. Kumar, and Saswati Chakraborty (2019). "Bioremediation of wastewater from automobile service station in anoxic-aerobic sequential reactors and microbial analysis." *Chemical Engineering Journal*, 361:982-989. doi: 10.1016/j.cej.2018.12.164.

[6] Taghipour, Shabnam, Hosseini S. Mossa, and Ataie-Ashtiani, Behzad. 2019. "Engineering Nanomaterials for Water and Wastewater Treatment: Review of Classifications, Properties and Applications." *New Journal of Chemistry*. doi: https://doi.org/10.1039/C9NJ00157C.

[7] Amir-Heidari, Payam, Arneborg, Lars, Lindgren, J. Fredrik, Andreas Lindhe, Rosén, Lars, Raie, Mohammad, Axell, Lars, and Hassellöv, Ida-Maja. 2019. "A state-of-the-art model for spatial and stochastic oil spill risk assessment: A case study of oil spill from a shipwreck." *Environment International*, 126:309-320. doi: 10.1016/j.envint.2019.02.037.

[8] Amir-Heidari, Payam and Raie, Mohammad 2019. "Response planning for accidental oil spills in Persian Gulf: A decision support system (DSS) based on consequence modeling." *Marine Pollution Bulletin*, 140:116-128. doi: 10.1016/j.marpolbul.2018.12.053.

[9] Amir-Heidari, Payam, and Raie, Mohammad. 2018. "Probabilistic risk assessment of oil spill from offshore oil wells in Persian Gulf." *Marine Pollution Bulletin*, 136: 291-299. doi: 10.1016/j.marpolbul.2018.07.068.

[10] Amir-Heidari, Payam, Farahani, Hadi, and Ebrahemzadih, Mehrzad. 2015. "Risk assessment of oil and gas well drilling activities in Iran– a case study: human factors." *International Journal of Occupational Safety and Ergonomics*, 21(3):276-283. doi: 10.1080/10803548.2015.1085162.

[11] Alavi, Nadali, Payam Amir-Heidari, Roza Azadi, and Ali-Akbar Babaei. 2013. "Effluent quality of ammonia unit in Razi petrochemical complex." *Journal of Advances in Environmental Health Research*, 1(1):15-20. doi: 10.22102/jaehr.2013.40120.

[12] Amir-Heidari, Payam, Maknoon, Reza, Taheri, Bahram, and Bazyari, Mahdieh. 2016. "Identification of strategies to reduce accidents and losses in drilling industry by comprehensive HSE risk assessment—A case study in Iranian drilling industry." *Journal of Loss Prevention in the Process Industries*, 44:405-413. Doi: 10.1016/j.jlp. 2016.09. 015.

[13] Khorsandi, Hassan, Teymori, Maryam, Aghapour, A. Ahmad, Jafari, S. Javad, Taghipour, Shabnam, and Bargeshadi, Rogaieh. 2019. "Photodegradation of ceftriaxone in aqueous solution by using UVC and UVC/H_2O_2 oxidation processes." *Applied Water Science*, 9(4):81. doi: 10.1007/s13201-019-0964-2.

[14] Amir-Heidari, P., Ebrahemzadih, M., Farahani, H. and Khoubi, J. (2014). Quantitative risk assessment in Iran's natural gas distribution network. *Open Journal of Safety Science and Technology*, 4(01), 59. doi: 10.4236/ojsst.2014.41008.

[15] Yu, Li, Han, Mei, and He, Fang. 2017. "A review of treating oily wastewater." *Arabian Journal of Chemistry*, 10:S1913-S1922. doi: 10.1016/j.arabjc.2013.07.020.

[16] Nagappan, Subbiah, Phinney, David, and Heldman, Dennis. 2018. "Management of Waste Streams from Dairy Manufacturing Operations Using Membrane Filtration and Dissolved Air Flotation." *Applied Sciences*, 8(12):2694. doi: 10.3390/app8122694.

[17] Johnson, Olufemi A., and Affam A. Chioma. 2018. "Petroleum sludge treatment and disposal: A review." *Environmental Engineering Research*, 24(2): 191-201. doi: 10.4491/eer.2018.134.

[18] Santo, Carlos E., Vilar J. P. Vítor, Botelho, Cidália M. S., Bhatnagar, Amit, Kumar, Eva, and Boaventura, Rui, A. R. 2012. "Optimization of coagulation–flocculation and flotation parameters for the treatment of a petroleum refinery effluent from a Portuguese plant." *Chemical Engineering Journal*, 183:117-123. doi: 10.1016/j.cej.2011.12.041.

[19] Abbassi, Bassim, and Taylor Livingstone. 2018. "A Comparative Review and Multi-criteria Analysis of Petroleum Refinery Wastewater Treatment Technologies." *Environmental Research, Engineering and Management*, 74(4):66-78. doi: 10.5755/j01.erem.74.4.21428.

[20] Schneider, EEa, Cerqueira, A. C. F. P., and Dezotti, Ma. 2011. "MBBR evaluation for oil refinery wastewater treatment, with post-ozonation and BAC, for wastewater reuse." *Water Science and Technology*, 63(1):143-148. doi: 10.2166/wst.2011.024.

[21] Varjani, Sunita, Joshi, Rutu, Srivastava, Vijay K., Ngo, H. Hao, and Wenshan Guo. 2019. "Treatment of wastewater from petroleum industry: current practices and perspectives." *Environmental Science and Pollution Research*, 1-9. doi: 10.1007/s11356-019-04725-x.

[22] Mulder, R., Vereijken, T. L. F. M., Frijters, C. M. T. J., and Vellinga, S. H. J. 2001. "Future perspectives in bioreactor development." *Water Science and Technology*, 44(8), 27-32. doi: 10.2166/wst.2001.0457.

[23] Nancharaiah, Y. V., and Kumar Reddy, G. K. 2018. "Aerobic granular sludge technology: mechanisms of granulation and biotechnological applications." *Bioresource Technology*, 247: 1128-1143. doi: 10.1016/j.biortech.2017.09.131.

[24] de Sousa Rollemberg, S. Luiz, Barros, R. M. Antônio, Firmino P. I. Milen, and dos Santos A. Bezerra. 2018. "Aerobic granular sludge: Cultivation parameters and removal mechanisms." *Bioresource Technology*. 270:678-688. doi: 10.1016/j.biortech.2018.08.130.

[25] Taghipour, Shabnam. 2015. *Study of GSBAR capability in petroleum wastewater treatment*. M.Sc. Thesis, Tarbiat Modares University.

[26] Szabó, Enikö. 2017. *Composition and dynamics of the bacterial community in aerobic granular sludge reactors*. PhD diss., Chalmers University of Technology.

[27] Tay, J. H., Tay, S. T. L., Liu, Y., Show, K. Y. and Ivanov, V. 2006. *Biogranulation technologies for wastewater treatment: Microbial granules* (Vol. 6). Elsevier.

[28] Moustafa, Tamara. 2014. *Aerobic granular sludge–study of applications for industrial and domestic wastewater.* M.Sc. Thesis, Chalmers University of Technology.

[29] Kong, Yunhua, Liu, Yong-Qiang, Tay, Joo-Hwa, Wong, Fook-Sin, and Zhu, Jianrong. 2009. "Aerobic granulation in sequencing batch reactors with different reactor height/diameter ratios." *Enzyme and Microbial Technology*, 45(5):379-383. doi: 10.1016/j.enzmictec. 2009.06.014.

[30] Taghipour, Shabnam, Ayati, Bita, and Razaei, Mina. 2017. "Study of the SBAR performance in COD removal of Petroleum and MTBE." *Modares Civil Engineering Journal*, 17(4):17-27. http://mjms. modares.ac.ir/article-16-7139-en.html.

[31] Ozturk, Arzu, Aygun, Ahmet, and Nas, Bilgehan. 2019. "Application of sequencing batch biofilm reactor (SBBR) in dairy wastewater treatment." *Korean Journal of Chemical Engineering*, 36(2): 248-254. doi: 10.1007/s11814-018-0198-2.

[32] Nancharaiah, Y. V., Sarvajith, M., and Lens, P. N. L. 2018. "Selenite reduction and ammoniacal nitrogen removal in an aerobic granular sludge sequencing batch reactor." *Water Research*, 131:131-141. doi: 10.1016/j.watres.2017.12.028.

[33] Hamza, R. Ahmed, Iorhemen, O. Terna, Zaghloul, Mohamed, S., and Tay, J. Hwa. 2018. "Rapid formation and characterization of aerobic granules in pilot-scale sequential batch reactor for high-strength organic wastewater treatment." *Journal of Water Process Engineering*, 22:27-33. doi: 10.1016/j.jwpe.2018.01.002.

[34] Speight, James G. 2014. *The chemistry and technology of petroleum.* CRC press.

[35] Fingas, Merv, and Ben Fieldhouse. (2014). Water-in-oil emulsions: formation and prediction. *Handbook of oil spill science and technology*, 225.

[36] Malakahmad, Amirhossein, Hasani, Amirhesam, Eisakhani, Mahdieh, and Isa, Mohamed H. 2011. "Sequencing Batch Reactor (SBR) for the removal of Hg^{2+} and Cd^{2+} from synthetic

petrochemical factory wastewater." *Journal of Hazardous Materials*, 191(1-3): 118-125. doi: 10.1016/j.jhazmat.2011.04.045.

[37] Yavuz, Yusuf, Koparal, A. Savaş, and Öğütveren Ü. Bakır. 2010. "Treatment of petroleum refinery wastewater by electrochemical methods." *Desalination*, 258(1-3): 201-205. doi: 10.1016/j. desal. 2010.03.013.

[38] Jafarinejad, S. (2016). *Petroleum waste treatment and pollution control*. First edition, Elsevier Inc., Butterworth-Heinemann, USA.

[39] Benyahia, Farid, Abdulkarim, Mohamed, Embaby, Ahmed, and Rao, Madduri. 2006. "Refinery wastewater treatment: A true technological challenge." Paper presented at the *The Seventh Annual UAE University Research Conference. UAE University*.

[40] Raut, Sangeeta, Sen, Sudip K. 2017. *Environmental engineering and safety*. Scientific Publishers.

[41] http:// cpcb.nic.in/ uploads/ Industry-Specific-Standards/ Effluent/ 03-petroleum_oil_refinery.pdf (G.S.R 186(E), dated 18th March, 2008) (Last accessed: 05.04.2019).

[42] Neisi, Abdolkazem, Afshin, Shirin, Rashtbari, Yousef, Babaei, A. Akbar, Khaniabadi Yusef O., Anvar Asadi, Shirmardi, Mohammad, and Vosoughi, Mehdi. 2018. "Efficiency of sequencing batch reactor for removal of organic matter in the effluent of petroleum wastewater." *Data in Brief*, 19:2041-2046. doi: 10.1016/j.dib.2018. 06.094.

[43] Shah, Maulin. 2014. "An application of sequencing batch reactors in microbial degradation of benzene, toluene, & xylene under anoxic and micro aerobic condition." *Applied and Environmental Microbiology*, 2:231-236. doi: 10.12691/jaem-2-5-5.

[44] Ma, G. and Love, N. G. 2001. "Creating anoxic and microaerobic conditions in sequencing batch reactors treating volatile BTX compounds." *Water Science and Technology*, *43*(3):275-282. doi: 10.2166/wst.2001.0147.

[45] Chiavola, A., Baciocchi, R. and Gavasci, R. 2010. "Biological treatment of PAH-contaminated sediments in a Sequencing Batch

Reactor." *Journal of Hazardous Materials*, 184(1-3): 97-104. doi: 10.1016/j.jhazmat.2010.08.010.

[46] Nikathirah, Yusoff, Ong, Soon-An, Ho, Li-Ngee, Wong, Yee-Shian, Saad Farah N. M., Khalik, WanFadhilah, and Lee, Sin-Li. 2019. "Performance of the hybrid growth sequencing batch reactor (HG-SBR) for biodegradation of phenol under various toxicity conditions." *Journal of Environmental Sciences*, 75:64-72. doi: 10.1016/j.jes.2018.03.001.

[47] Al-Khalid, Taghreed, and El-Naas. H. Muftah. 2017. "Organic Contaminants in Refinery Wastewater: Characterization and Novel Approaches for Biotreatment." *In Recent Insights in Petroleum Science and Engineering*. IntechOpen. doi: 10.5772/intechopen. 72206.

[48] Mustapha, Hassana, I., van Bruggen, J. J. A., and Lens, P. N. L. 2018. "Fate of heavy metals in vertical subsurface flow constructed wetlands treating secondary treated petroleum refinery wastewater in Kaduna, Nigeria." *International Journal of Phytoremediation*, 20(1):44-53. doi: 10.1080/15226514.2017.1337062.

[49] Sekine, Mutsumi, Akizuki, Shinichi, Kishi, Masatoshi, and Toda, Tatsuki. "Stable nitrification under sulfide supply in a sequencing batch reactor with a long fill period." *Journal of Water Process Engineering*, 25:190-194. doi: 10.1016/j.jwpe.2018.05.012.

[50] El-Sheekh, Mostafa, M., Metwally A. Metwally, Allam, Nanis G., and Hemdan Hany E. 2017. "Effect of Algal Cell Immobilization Technique on Sequencing Batch Reactors for Sewage Wastewater Treatment." *International Journal of Environmental Research*, 11(5-6): 603-611. doi: 10.1007/s41742-017-0053-z.

[51] Toh, S., Tay, J., Moy, B., Ivanov, V., and Tay, S. 2003. "Size-effect on the physical characteristics of the aerobic granule in a SBR." *Applied Microbiology and Biotechnology*, 60(6):687-695. doi: 10.1007/s00253-002-1145-y.

[52] Liu, Changli, Liu, Di, Qi, Yingjie, Zhang, Ying, Liu, Xi, and Zhao, Min. 2016. "The effect of anaerobic–aerobic and feast–famine cultivation pattern on bacterial diversity during poly-β-

hydroxybutyrate production from domestic sewage sludge." *Environmental Science and Pollution Research*, 23(13):12966-12975. doi: 10.1007/s11356-016-6345-6.

[53] Xing, Bao-Shan, Guo, Qiong, Jiang, Xiao-Yan, Chen, Qian-Qian, He, Miao-Miao, Wu, Li-Min, and Jin, Ren-Cun. 2016. "Long-term starvation and subsequent reactivation of anaerobic ammonium oxidation (anammox) granules." *Chemical Engineering Journal*, 287:575-584. doi: 10.1016/j.cej.2015.11.090.

[54] Li, Xiaoyan, Jin, Zijing, Qian, Yongyu, Cui, Daizong, Chen, Xiguang, and Zhao, Min. 2017. "Production of poly-β-hydroxybutyrate by activated sludge in sequencing batch reactor under aerobic conditions." *Journal of Wuhan University of Technology-Mater. Sci. Ed.*, 32(3):733-738. doi: 10.1007/s11595-017-1660-4.

[55] Li, Yiyu, Yu, Tao, Kang, Da, Shan, Xiaoyu, Zheng, Ping, Hu, Zhiqiang, Ding, Aqiang, Wang, Ru, and Zhang, Meng. 2019. "Sources of anammox granular sludge and their sustainability in treating low-strength wastewater." *Chemosphere*. 226:229-237. doi: 10.1016/j.chemosphere.2019.03.049.

[56] Zhang, Yuanyuan, Kuroda, Masashi, Arai, Shunsuke, Kato, Fumitaka, Inoue, Daisuke, and Ike, Michihiko. 2019. "Biological treatment of selenate-containing saline wastewater by activated sludge under oxygen-limiting conditions." *Water Research*, 154:327-335. doi: 10.1016/j.watres.2019.01.059.

[57] Buzzini, Andréa P., Sakamoto Isabel K., Varesche, M. B., and Pires, Eduardo C. 2006. "Evaluation of the microbial diversity in an UASB reactor treating wastewater from an unbleached pulp plant." *Process Biochemistry*, 41(1):168-176. doi: 10.1016/j.procbio.2005.06.009.

[58] Feng, Qian, Cao, Jia-sun, Chen, Li-Na, Guo, Chen-Yuan, Tan, Jun-yi, and Xu, Hui-lian. 2012. "Effect of carbon source on biological nitrogen removal of aerobic granules in sequencing batch reactors." *Journal of Food, Agriculture & Environment*, 10(2):1110-1113.

[59] Wang, Lawrence K., Shammas, Nazih K., and Hung, Yung-Tse, eds. 2010. *Advanced Biological Treatment Processes* (Vol. 9). Springer Science & Business Media.

[60] Boztoprak, Halime, Özbay, Yüksel, Güçlü, Dünyamin, and Küçükhemek, Murat. 2016. "Prediction of sludge volume index bulking using image analysis and neural network at a full-scale activated sludge plant." *Desalination and Water Treatment*, 57(37):17195-17205. doi: 10.1080/19443994.2015.1085909.

[61] Pronk, M., De Kreuk, M. K., Bruin, B. De, Kamminga, P., Kleerebezem, R. van, and Van Loosdrecht, M. C. M. 2015. "Full scale performance of the aerobic granular sludge process for sewage treatment." *Water Research*, 84:207-217. doi: 10.1016/j.watres. 2015.07.011.

[62] Kang, Abbass Jafari, and Yuan, Qiuyan. 2017. "Long-term stability and nutrient removal efficiency of aerobic granules at low organic loads."*Bioresource Technology*, 234:336-342. doi: 10.1016/j. biortech.2017.03.057.

[63] Cofré, C., Campos, J. L., Valenzuela-Heredia, D., Pavissich, J. P., Camus, N., Belmonte, M., Pedrouso, A., Carrera, P., Mosquera-Corral, A., and Val del Río, A. 2018. "Novel system configuration with activated sludge like-geometry to develop aerobic granular biomass under continuous flow." *Bioresource Technology*, 267:778-781. doi: 10.1016/j.biortech.2018.07.146.

[64] Tassew, Fasil A., Bergland, W. Hennie, Dinamarca, Carlos, and Bakke, Rune. 2019. "Settling velocity and size distribution measurement of anaerobic granular sludge using microscopic image analysis." *Journal of Microbiological Methods*, 159:81-90. doi: 10.1016/j.mimet.2019.02.013.

[65] Wang, Zhiyao, and Zheng, Ping. 2017. "Predicting settling performance of ANAMMOX granular sludge based on fractal dimensions." *Water Research*, 120:222-228. doi: 10.1016/j.watres. 2017.03.056.

[66] Deen, W. M. 2016. *Introduction to Chemical Engineering Fluid Mechanics*. Cambridge University Press.

[67] Rosman, N. Hasyimah, Anuar N. Aznah, Chelliapan. Shreeshivadasan, Din. Mohd F. M., and Ujang, Zaini. 2014. "Characteristics and performance of aerobic granular sludge treating rubber wastewater at different hydraulic retention time."*Bioresource Technology*, 161: 155-161. doi: 10.1016/j.biortech.2014.03.047.
[68] Long, Bei, Yang, Chang-zhu, Pu, Wen-hong, Yang, Jia-kuan, Jiang, Guo-sheng, Dan, Jing-feng, Li, Chun-yang, and Liu, Fu-biao. 2014. "Rapid cultivation of aerobic granular sludge in a pilot scale sequencing batch reactor." *Bioresource Technology*, 166:57-63. doi: 10.1016/j.biortech.2014.05.039.
[69] Dahalan, Farrah A., Najib Mohamed Z. M., Salim, Mohd R., and Ujang Zaini. 2015. "Characteristics of developed granules containing phototrophic aerobic bacteria for minimizing carbon dioxide emission." *International Biodeterioration & Biodegradation*, 102:15-23. doi: 10.1016/j.ibiod.2015.04.010.
[70] Dai, Yajie, Jiang, Yixin, and Su, Haijia. 2015. "Influence of an aniline supplement on the stability of aerobic granular sludge." *Journal of Environmental Management*, 162:115-122. doi: 10.1016/j.jenvman.2015.05.017.

BIOGRAPHICAL SKETCH

Bita Ayati

Affiliation: Civil & Environmental Engineering Faculty, Tarbiat Modares University, Tehran, Iran

Education: Environmental Engineering

Business Address: Civil & Environmental Engineering Faculty, Tarbiat Modares University, P.O. Box 14115- 397, Tehran, Iran

Research and Professional Experience: Water treatment, Sanitary and industrial wastewater treatment

Professional Appointments: Assoc. Prof.

Publications from the Last 3 Years:

1. Hooshmand, S., B. Ayati, Optimization of Effective Parameters in Removal of Acid Blue 25 by Fenton Process with the Aim of Reducing Iron Consumption, Accepted for publication in *Journal of Environmental Science and Technology* (JEST) (in Persian).
2. Zanganeh Ranjbar, P., B. Ayati, H. Ganjidoust, Kinetic study on photocatalytic degradation of Acid Orange 52 in a baffled reactor using TiO_2 nanoparticles, *Journal of Environmental Sciences*, Vol. 79, pp. 213-224, May 2019, https://doi.org/10.1016/j.jes.2018.06.012.
3. Arbab, P., B. Ayati, M. R. Ansari, Application of Hydrodynamic Cavitation Process for Dye Removal and Optimization based on Energy Consumption, *J. of Environmental Sciences*, Vol. 16, No. 3, pp. 119- 134, Autumn 2018.
4. Arbab, P., B. Ayati, M. R. Ansari, Reducing the Use of Nanotitanium Dioxide by *Switching from Single Photocatalysis to Combined Photocatalysis-Cavitation in Dye Elimination, Process Safety and Environmental Protection*, Vol. 121, pp. 87-93, January 2019, https://doi.org/10.1016/j.psep.2018.10.012.
5. Irankhah, S., A. Abdi Ali, M.R. Soudi, S. Gharavi, B. Ayati, Highly efficient phenol degradation in a batch moving bed biofilm reactor: benefiting from biofilm-enhancing bacteria, *Journal of Microbiology and Biotechnology*, Vol. 34, No. 164, Online: October 2018, https://doi.org/10.1007/s11274-018-2543-3.
6. Nazari, M., B. Ayati, Investigation of Anionic Surfactant Removal Using Unipolar Electro-Flotation and Electro-Coagulation, *J. of Water and Wastewater,* Vol. 29, Issue 3, July and August 2018, pp. 54-65, doi: 10.22093/wwj.2017.72005.2316 (in Persian).

7. Arabyarmohammadi, H., A. K. Darban, M. Abdollahy, R. Yong, B. Ayati, A. Zirakjou, S. E. A. T. M. van der Zee, Utilization of a Novel Chitosan/Clay/Biochar Nanobiocomposite for Immobilization of Heavy Metals in Acid Soil Environment, *Journal of Polymers and the Environment,* Vol. 26, Issue 5, pp. 2107–2119, May 2018, DOI 10.1007/s10924-017-1102-6.
8. Arabyarmohammadi, H., A. Khodadadi Darban, S. E. A. T. M. van der Zee, M. Abdollahy, B. Ayati, *Fractionation and Leaching of Heavy Metals in Soils Amended with a New Biochar Nanocomposite, Environmental Science and Pollution Research*, Vol. 25, Issue 7, pp. 6826–6837, March 2018, https://doi.org/10.1007/s11356-017-0976-0.
9. Ghobadian, S., H. Ganjidoust, B. Ayati, N. Soltani, The Innovative Engineered Photobioreactor to Optimize the Amount of Microalgae Spirulina Biomass, *Nova Biologica Reperta*, Vol. 5, Issue 1, pp. 13-25, 2018.
10. Ahangar, M. A., H. Ganjidoust, B. Ayati, Optimization of Parameters Depended on the Electrode in the Dye Treatment by Use of the Electro- Coagulation- Flotation Process, *Sharif J. of Science and Technology*, Vol. 34.2, Issue 1 & 2, pp. 25- 33, Spring 2018 (in Persian).
11. Zanganeh Ranjbar, P., B. Ayati, H. Ganjidoust, Modeling of a Photocatalytic Baffled Reactor to Degrade Colored Wastewater using Response Surface Methodology, *Modares Civil Engineering Journal*, Vol. 18, No. 1, pp. 113-122, Spring 2018 (in Persian).
12. Ahangarnokolaei, M. A., H. Ganjidoust, B. Ayati, Optimization of parameters of Electrocoagulation/ Flotation Process for Removal of Acid Red 14 with Mesh Stainless Steel Electrodes, Accepted for publication in *Journal of Water Reuse and Desalination*, doi: 10.2166/wrd.2017.091.
13. Ahangar, M. A., H. Ganjidoust, B. Ayati, Colored Wastewater Treatment Using Electro-coagulation-flotation Method with Mesh Stainless Steel Electrode, *Journal of Environmental Sciences*, Vol. 43, Issue 2, Summer 2017, pp. 195- 206 (in Persian).

14. Sayyahzadeh, A. H., H. Ganjidoust, B. Ayati, Removal Mechanisms Involved in the Petroleum Refinery Wastewater Treatment by MBBR System, *J. of Water and Wastewater*, Volume 28, Issue 3, pp. 87- 96, DOI:10.22093/wwj.2016.40798 (in Persian).
15. Taghipour, S., B. Ayati, Cultivation of aerobic granules through synthetic petroleum wastewater treatment in a cyclic aerobic granular reactor, *Journal of Desalination and Water Treatment*, Vol. 76, pp. 134–142, May 2017, doi:10.5004/dwt.2017.20779.
16. Soroush, F., H. Ganjidoust, B. Ayati, Removal of Petroleum Hydrocarbons from Contaminated Waters using a Solar Photocatalytic Process, *Journal of Ferdowsi Civil Engineering*, Vol. 29, No. 1, pp. 37-48, Spring 2017, doi: 10.22067/civil.v29i1.42661 (in Persian).
17. Zoqi, M. J., H. Ganjidoust, N. Mokhtarani, B. Ayati, Effect of Inorganic Material and Non-uniform Electrokinetic on Solidification/Stabilization of Lead, Zinc and Arsenic, *Sharif J. of Science and Technology*, Vol. 33.2, Issue 1.2 - pp. 79- 89, Spring 2017 (in Persian).
18. Amiri, H., B. Ayati, H. Ganjidoust, Textile Dye Removal using a Photocatalytic Cascade Disc Reactor Coated by ZnO Nanoparticles: The Effects of Hydraulic Parameters, *ASCE's Journal of Environmental Engineering*, Vol. 142, Issue 6, June 2017, https://doi.org/10.1061/(ASCE)EE.1943-7870.0001092.
19. Larimi, S. N., B. Ayati, An investigation on Removal Efficiency of Direct Blue 71 by ZnO Nano-Particles and Activated Carbon Produced from Agricultural Wastes, *Iranica Journal of Energy & Environment,* Vol. 8, No.2, pp. 118- 126, Spring 2017.
20. Minaei Zangi, Z., H. Ganjidoust, B. Ayati, Analysis of Photocatalytic Degradation of Azo Dyes under Sunlight with Response Surface Method, *Journal of Desalination & Water Treatment*, Vol. 63, pp. 262- 274, Feb. 2017, doi: 10.5004/dwt.2017.20164.
21. Amiri, H., B. Ayati, H. Ganjidoust, Mass Transfer Phenomenon in Photocatalytic Cascade Disc Reactor: Effects of Artificial Roughness and Flow Rate, *Chemical Engineering & Processing: Process*

Intensification, Vol. 116, pp. 48- 59, June 2017, doi: 10.1016/j.cep. 2017.03.004.

22. Rahimi, M. G., B. Ayati, A. Rezaei, Optimization of Reaction Parameters for the Sonophotocatalytic Degradation of Hydroquinone, *Research on Chemical Intermediates*, Vol. 43, Issue 3, pp. 1935–1956, March 2017, doi: 10.1007/s11164-016-2740-3.
23. Ayati, B., M. Rezaei, Modified Sequencing Batch Airlift Reactor Capability in MTBE Removal, *Iranica Journal of Energy & Environment,* Vol. 8, No.1, pp. 26- 30, 2017 (Research Note).
24. Niksefat, M., B. Ayati, Synthesis of Bimetallic Nanoparticles Fe-Ni and Investigation of their Performance in AR14 Dye Removal from Aqueous Solutions, *Sharif J. of Science and Technology*, Vol. 32.2, Issue 4.2, pp. 33-44, Winter 2017 (in Persian).
25. Rahimi, M. G., A. Khodadadi Darban, B. Ayati, Photolysis System Performance in Petroleum Hydrocarbons Removal from Wastewater and its Modeling, *Journal of Ferdowsi Civil Engineering*, Vol. 28, No. 2, pp. 1- 8, Winter 2017 (in Persian).
26. Nazirian, P., Bi. Ayati, H. Ganjidoust, Comparing the Capability of Photocatalyst Nano Zinc Oxide Process by Two Slurry and Immobilized Methods in Dye Acid Orange 7 Removal, *J. of Environmental Studies,* Vol. 42, Issue 4, pp. 855- 867, Winter 2017 (in Persian).
27. Ayati, B., M. Razaei, SBAR Capability in Methyl Tert Butyl Ether Removal: Characteristics of the Bio-granules, *J. of Water and Wastewater*, Vol. 27, Issue 6, pp. 50- 58, December 2016 (in Persian).
28. Mousavi, S. M. S., B. Ayati, H. Ganjidoust, Phenol Removal and Bio-electricity Generation using a Single-chamber Microbial Fuel Cell in Saline and Increased- Temperature Condition, *Journal of Energy Sources, Part A: Recovery, Utilization, and Environmental Effects*, Volume 38, Issue 22, pp. 3300-3307, Oct. 2016, http://dx.doi.org/10.1080/15567036.2016.1156196.
29. Kalhor, H., H. Ganjidoust, B. Ayati, Simultaneous Removal of Salinity and Organic Loading Rate using Phytoremediation, *J. of*

Environmental Studies, Vol. 42, Issue 3, pp. 531- 550, Autumn 2016 (in Persian).

30. Niksefat, M., B. Ayati, Reduction and Reuse of Fe-Ni Bimetallic Nanoparticles Oxide and Evaluating Its Ability In *Acid Red 14 Removal, Environmental Progress & Sustainable Energy*, Vol. 35, No. 6, pp. 1646- 1656, Nov. 2016, doi: 10.1002/ep.12407.
31. Mokhtarani, N., S. Khodabakhahi, B. Ayati, Optimization of Photocatalytic Post-Treatment of Composting Leachate using UV/TiO_2, *Desalination and Water Treatment*, Vol. 57, Issue 47, pp. 22232-22243, 2016, doi:10.1080/19443994.2015.1130652.
32. Ghalebizadeh, M., B. Ayati, Investigation of Dye Removal Efficiency of the Photoelectrocatalytic System Using Graphite and Stainless Steel as Electrodes, *J. of Water and Wastewater*, Vol. 27, Issue 4, pp. 26-35, Sept. and Oct. 2016 (in Persian).
33. Ghalebizadeh, M., B. Ayati, Solar Photoelectrocatalytic Degradation of Acid Orange 7 with ZnO/TiO_2 Nanocomposite Coated on Stainless Steel Electrode, *Process Safety and Environmental Protection Journal*, Vol. 103, pp. 192- 202, Sept. 2016, doi: 10.1016 /j.psep.2016.07.009.
34. Rahimi, S., B. Ayati, A. Rezaei, Kinetic Modeling and Determination Role of Sono/Photo Nanocatalyst-Generated Radical Species on Degradation of Hydroquinone in Aqueous Solution, *Environmental Science and Pollution Research*, Vol. 23, Issue 2, pp. 12185- 12198, · June 2016, doi: 10.1007/s11356-016-6408-8.
35. Zoghi, M. J., H. Ganjidoust, N. Mokhtarani, B. Ayati, Solidification Optimization of Electroplating Sludge, *Journal of Environmental Engineering and Science*, Vol. 11, Issue JS2, pp. 33- 43, June 2016, DOI: http://dx.doi.org/10.1680/jenes.16.00005.
36. Niksefat, M., A. Hooshmandfar, B. Ayati, Synthesis of Bimetallic Fe-Ni Nano-Particles and using them to Improve Electrochemical Treatment of Wastewater, *Iranian Journal of Mining Engineering Journal*, Vol. 11, Issue 31, Autumn 2016, pp. 81-89 (in Persian).
37. Sayyahzadeh, A. H., H. Ganjidoust, B. Ayati, MBBR System Performance Improvement for Petroleum Hydrocarbon Removal

using Modified Media with Activated Carbon, *Water Science and Technology*, Vol. 73, No. 9, pp. 2275-83, May 2016, doi: 10.2166/wst.2016.013.

38. Asgari, R., B. Ayati, Role of Electron Scavengers in Direct Blue 71 Removal by Nano-TiO_2 Immobilized on Cementitious Bed, *Journal of Environmental Studies*, Vol. 42, Issue 1, pp. 1-17, Spring 2016 (in Persian).
39. Mohammadizad, M., H. Ganjidoust, B. Ayati, Determining the Optimum Conditions for Photocatalytic Dye Removal by Polyaniline/Graphene Nano-Composite Under Visible Light Irradiation, *Journal of Color Science and Technology*, Vol. 10, No. 1, pp. 31- 42, Spring 2016 (in Persian).
40. Hossini, H., A. Rezaei, B. Ayati, A. H. Mahvi, Optimizing Ammonia Volatilization by Air Stripping from Aquatic Solutions using Response Surface Methodology (RSM), *Desalination and Water Treatment*, Vol. 57, Issue 25, pp. 11765-11772, May 2016, doi: 10.1080/19443994.2015.1046946.
41. Ghalebizadeh, M., B. Ayati, Analysis of Degradation of Acid Orange 7 by Electro-Fenton Process with Graphite Cathode Coated by Carbon Nanotubes, *Modares Civil Engineering Journal*, Vol. 16, Issue 1, pp. 117-126, Winter 2016 (in Persian).
42. Niksefat, M., B. Ayati, Determine the Optimum Conditions for Acid Red 14 Removal by Iron Nanoparticles Modified with Nickel in a Slurry System, *J. of Water and Wastewater*, Vol. 27, Issue 1, pp. 29-39, March and April 2016 (in Persian).
43. Asgari, R., B. Ayati, Scavengers Effect on Accelerating Photocatalytic Removal of Direct Blue 71 Dye with Nano TiO_2 Immobilized on Cementitious Bed, *Sharif J. of Science and Technology*, Vol. 31.2, Issue 4.2, pp. 25- 35, Winter 2016 (in Persian).
44. Ghalebizadeh, M., B. Ayati, Analysis of Effective Parameters on Electro Fenton System with Stainless Steel and Graphite Electrodes for Removing Acid Orange 7, *J. of Environmental Studies*, Vol. 41, Issue 4, pp. 855-865, Winter 2016 (in Persian).

45. Taherzadeh, M., H. Ganjidoust, B. Ayati, Experimental Improvement Ways for Hydraulic Performance of Lab- Scale Stabilization Pond, *Iranian Chemical Engineering Journal*, Vol. 14, No. 82, pp. 30- 38, Winter 2016 (in Persian).
46. Hossini, H., A. Rezaee, B. Ayati, A. H. Mahvi, Off-gas Treatment of Ammonia using a Diffused Air Stripper: A Kinetic Study, *Health Scope*, Vol. 5, No. 1, pp. 26479-26485, February 2016, doi: 10.17795/jhealthscope-26479.
47. Raeisi, S. M., M. Tabatabaei, B. Ayati, A. Ghafari, S. Haghigi Mood, A Novel Combined Pretreatment Method for Rice Straw Using Optimized EMIM [Ac] and Mild NaOH, *Waste and Biomass Valorization*, Vol. 7, Issue 1, pp. 97- 107, February 2016.
48. Hooshmandfar, B. Ayati, A. Khodadadi Darban, Optimization of Material and Energy Consumption for Removal of Acid Red 14 by Simultaneous Electro-coagulation and Electro-flotation, *Water Science & Technology*, Vol. 73, No. 1, pp. 192-202, January 2016, doi:10.2166/wst.2015.477.

Shabnam Taghipour

Affiliation: Civil & Environmental Engineering, Tarbiat Modares University, Tehran, Iran.

Education: Msc in Civil and Environmental Engineering

Research and Professional Experience:

- Application of Nanotechnology in water, wastewater and groundwater remediation
- Application of Advanced Oxidation Processes (AOPs) in water, wastewater and groundwater remediation
- Experimental modeling of contaminant transport in groundwater

- Application of biological methods in water and wastewater treatment
- Teaching Assistant for courses:
  - Fluid Mechanics (Teaching Assistant for Dr. Ataie Ashtiani, Sharif University of Technology, Iran).
  - Environmental engineering (Teaching Assistant for Dr. Danesh Yazdi, Sharif University of Technology, Iran).
  - Open Channel Hydraulics (Teaching Assistant for Dr. Danesh Yazdi, Sharif University of Technology, Iran).
- Research scholar in two projects:
  - Studying the photocatalytic efficiency of zinc oxide nanoparticles in diclofenac degradation from aqueous environments (Project Director: Dr. Javad Jafari), Urmia University of Medical Sciences, Iran, (2017).
  - Investigating the efficiency of UV/H2O2 process in removal of ceftriaxone antibiotic from aqueous environments (Project Director: Dr. Javad Jafari), Urmia University of Medical Sciences, Iran, (2017).
- Teacher in workshops:
  - Urban Waste Management and Principle Disposal of Non-recyclable Solid and Semi-solid Wastes (20 hours, 2016).
  - Principles of disposal and degradation of environmental wastewater, goals and related risks (25 hours, 2016).
  - Infectious and hospital wastes in metropolises-How to work and design Waste incinerators (16 hours, 2016).

Honors:

- Best paper, International Conf. on Environmental Science, Engineering & Technology (CECET 2015), Tehran, Iran, (2015).
- Best student of Athlete Students Choice of Civil Engineering Department in Sharif University of Technology, Tehran, Iran, (2018).

Publications from the Last 3 Years:

Khorsandi H., Teimouri M., Aghapour AA., Jafari S. J., Taghipour S., Bargshadi R. Photodegradation of ceftriaxone in aqueous solutions by using UVC and UVC/H_2O_2 oxidation processes, *Applied Water Science*. 2019, 9(4):81.

Taghipour, Shabnam, and Ayati, Bita "Cultivation of aerobic granules through synthetic petroleum wastewater treatment in a cyclic aerobic granular reactor." *Journal of Desalination and Water Treatment* 76 (2017): 134-142.

Taghipour, Shabnam, Bita Ayati, and Mina Razaei. "Study of the SBAR performance in COD removal of Petroleum and MTBE." *Modares Civil Engineering journal* 17, no. 4 (2017): 17-27 (In persian).

Taghipour, Shabnam, Hosseini, Seiyed Mossa, and Ataie-Ashtiani, Behzad. "Engineering Nanomaterials for Water and Wastewater Treatment: Review of Classifications, Properties and Applications." *New Journal of Chemistry*. 2019, 43:7902-7927 (DOI: https://doi.org/10.1039/C9NJ00157C).

Taghipour, Shabnam, Khodadadi, A., *Principles of transport and diffusion of contaminants in surface waters*. Fadak Publication, Tehran, Iran (ISBN: 978-600-160-283-2), 2017 (In persian).

In: Wastewater Treatment
Editor: Adriana Magdalena
ISBN: 978-1-53616-370-4

Chapter 6

IMPROVING BIOMETHANE PRODUCTION BY THE ANAEROBIC CO-DIGESTION OF AGRO-INDUSTRIAL WASTES (VINASSE, WHEY AND GLYCERIN) USING SEQUENCING BATCH BIOFILM REACTORS

R. Albanez, G. Lovato, J. N. Albuquerque, S. P. Sousa, S. M. Ratusznei and J. A. D. Rodrigues*
Mauá School of Engineering, Mauá Institute of Technology (EEM/IMT), Praça Mauá 1, São Caetano do Sul, SP, Brazil

ABSTRACT

Anaerobic digestion is a promising technology for wastewater treatment as it allows the recovery of energy from biogas (hydrogen and/or methane) generated as a final product of the process. In addition to its low-cost operation, it significantly reduces the polluting load of the liquid residue. However, biogas production via anaerobic digestion

* Corresponding Author's E-mail: rodrigues@maua.br.

depends greatly on substrate composition, such as fraction of organic matter, as well as on the amount of nutrients available for microorganisms. In this context, anaerobic co-digestion, consisting of the simultaneous anaerobic digestion of multiple organic residues in a single digester, is an interesting technique to improve organic waste digestion and its conversion to biogas.

The agro-industrial wastes vinasse (sugarcane stillage), whey and glycerin usually present medium or low process yields when digested alone, making their co-digestion with other residues, produced in large quantities and with complementary characteristics, an attractive option. Therefore, a mini review has been performed regarding studies that dealt with the co-digestion of these wastes (vinasse with whey, vinasse with glycerin and whey with glycerin) for methane production.

In these co-digestion processes an anaerobic sequencing batch (or fed-batch) biofilm reactor (AnSBBR) was used, which is one of the several high-rate configurations used as an alternative to continuous systems. Advantages of the AnSBBR include better effluent control and simple operation consisting of three stages: feed, reaction and discharge.

AnSBBRs have been applied to the treatment of vinasse (bioethanol production), whey (dairy industry) and glycerin (biodiesel production) using various operational strategies: feeding mode, temperature, organic load, influent concentration and cycle length. Thus, this review presents an overview of the achievements of studies in which AnSBBRs have been used to co-digest agro-industrial wastes for the production of methane, with the focus on operational strategy and perspectives for energy estimations.

Keywords: methane, co-digestion, vinasse, whey, glycerin, AnSBBR

1. INTRODUCTION

The anaerobic production of methane from a large variety of biological residues has increased world-wide because of its environmental and economic benefits. Biogas can be locally produced from renewable sources and does not depend on imported oil or natural gas stocks. It reduces the amount of pollution by organic residues, which are responsible for the major part of the pollution of natural water bodies. Furthermore, methane has a combustion energy of 55.6 $kJ.g^{-1}$, which is 17% greater than the amount of energy in gasoline (47.5 $kJ.g^{-1}$) (Chandra et al., 2012; Lovato et al., 2015).

Anaerobic digestion is a biological treatment performed in the absence of oxygen to stabilize organic matter while producing biogas, a mixture consisting mainly of methane and carbon dioxide. Anaerobic digestion of single substrates (mono-digestion) presents some drawbacks related to substrate properties. These problems can be solved by adding a co-substrate in what has been recently called anaerobic co-digestion. This technique consists in mixing wastes with complementary features to increase digestion speed, efficiency and stability. Co-digestion may provide dilution of toxic substances, substrate nutrient balance, synergistic effects on microorganisms, increase in biodegradable organic matter content and greater methane yield per unit volume of the digester (Albanez et al., 2016b; Lovato et al., 2019a; Mata-Alvarez et al., 2014; Shah et al., 2015).

Agro-industrial wastes are produced as a result of agricultural activities. An agricultural establishment produces many types of wastes during its daily operations. These include liquid or solid wastes that result from agricultural practices, such as cattle manure, crop residues, pesticides, and fertilizers. Agricultural wastes have become an increasing concern in recent years, as they may cause significant environmental problems; however, they may also be used for several beneficial purposes, such as feedstock for energy production from anaerobic digestion (Merlin and Boileau, 2013). Brazil stands out for the large production of the agro-industrial wastes vinasse (sugar cane stillage, from the sugar and ethanol production process), glycerin from the biodiesel industry and cheese whey from the dairy industry.

Cheese whey, or milk whey, is the watery part of milk that is separated from the curd in the cheese-making process. It is the major by-product of dairy industries, and because of the high organic load, it has strong pollution potential, causing an excess of oxygen consumption when directly disposed of in water bodies. Generally, cheese whey presents high organic load, low alkalinity content, high nitrogen content and very high biodegradability. On a global scale, only 50% of all produced cheese whey is used in industries and, on average, 873 mL of whey is generated per liter of cow milk, resulting in approximately 5 million tons of whey per year (Carvalho et al., 2013; Lovato et al., 2017a; Panesar et al., 2007; Rico et al., 2015).

Glycerol is the major by-product of the biodiesel industry. In general, for every 100 kg of biodiesel produced, approximately 10 kg of crude glycerol is generated, and the global biodiesel market is estimated to reach 37 billion gallons by 2016 with an average annual growth of 42%. Crude glycerol, generated by homogeneous base-catalyzed transesterification, contains approximately 50-60% glycerol, 12-16% alkalis, 15-18% methyl esters, 8-12% methanol, and 2-3% water (Lovato et al., 2017a; Rivero et al., 2014).

Vinasse is the main residue of the sugar and ethanol industry, 14 liters of vinasse are generated per liter of produced ethanol. It is worth noting that in the 2017/2018 harvest 400 billion liters of vinasse were generated. Vinasse is a complex wastewater that can be highly damaging to the environment in which it is discarded because of its high organic content (22-45 $gCOD.L^{-1}$), low pH, high temperature and high ash content. Vinasse may be considered a substrate for biogas production via anaerobic treatment because of its organic content (carbon, macro and micronutrients), being a promising renewable biofuel (Albanez et al., 2016a; Ferraz Junior et al., 2014; Leite et al., 2008).

One possible reactor choice to treat these wastes is the anaerobic sequencing batch biofilm reactor with immobilized biomass (AnSBBR) which is a technological option for discontinuous operation as an alternative to the commonly used continuous operation. A typical cycle of the anaerobic reactor operated in batch and fed-batch mode consists of four steps: (i) feeding which can have variable filling time, defining the filling strategy as batch and/or fed-batch; (ii) treatment itself by means of the biotransformation of the wastewater constituents by the microorganisms; (iii) sedimentation, when the biomass is in granular form (ASBR), because when the biomass is immobilized on an inert support (AnSBBR), this step is not necessary; and (iv) discharge, with removal of the treated and clarified liquid (Albanez et al., 2016a; Zaiat et al., 2001). It represents an option for small and medium agro-industrial waste producers who can find in the sequencing batch operation a more flexible structure regarding the type of feeding strategy and influent concentration.

Thus, this mini review presents an overview of achievements reported in studies that used AnSBBRs to co-digest the agro-industrial wastes vinasse, whey and glycerin for methane production, focusing on operational strategy and perspectives for energy estimations.

2. Material and Methods

2.1. AnSBBR with Recirculation of the Liquid Phase

Figure 1 shows a scheme of the system used for the production of biomethane via anaerobic treatment from the co-digestion of cheese whey, glycerin and vinasse. The reactor, with a capacity of 1.5 L of liquid medium, in addition to the inert support and biomass, consisted of an acrylic cylindrical column with a height of 540 mm, an external diameter of 100 mm and a wall thickness of 3.5 mm with a total volume of 3.0 L. An automatic unit for feeding, discharging and recirculating the liquid phase was provided.

The inert support containing the immobilized biomass was placed between perforated stainless-steel plates, which divided the 540-mm high reactor into five parts to avoid bed compacting. A 20-mm compartment at the bottom of the reactor allowed enhanced distribution of the wastewater, preventing formation of preferential routes. At the upper part of the reactor, a 40-mm region functioned as a biogas collector (CH_4 and CO_2) (Lovato et al., 2016).

The recirculation unit comprised (i) a side reservoir containing 1.5 L of liquid medium, consisting of a cylindrical acrylic container with a height of 300 mm, an external diameter of 100 mm and a wall thickness of 3.5 mm (total capacity of 2.1 L) and (ii) a diaphragm pump with a maximum capacity of 30 $L.h^{-1}$. A 100-mL measuring cylinder was attached to the recirculation system for flow rate measurements. The total volume of the reaction medium (1.5 L contained in the reactor and 1.5 L in the parallel tank) was 3.0 L (Lovato et al., 2016).

Feeding and discharge were performed using diaphragm pumps equipped with automatic timers. Temperature was kept at 30±1°C by a heating system composed of resistance heaters and fans as well as a temperature sensor and controller (Lovato et al., 2016).

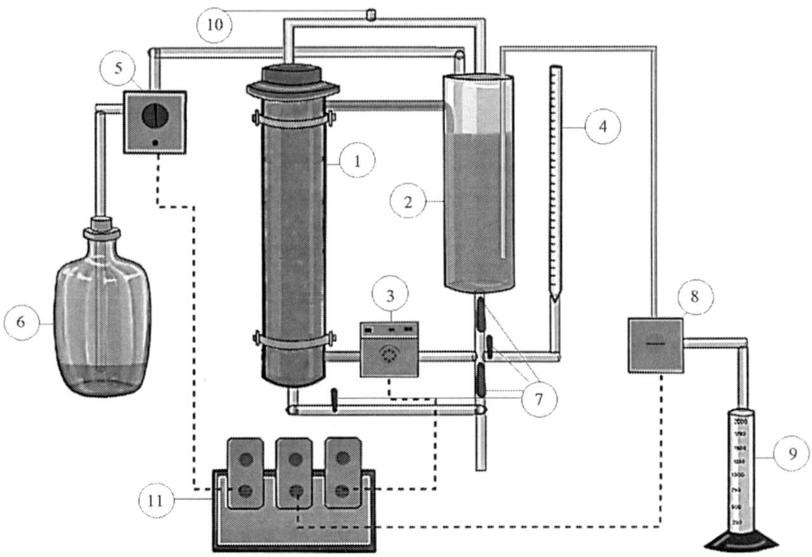

[Notation: 1 – Reactor with immobilized biomass; 2 – Side reservoir; 3 – Recycle pump; 4 – Flow meter; 5 – Feed pump; 6 – Wastewater reservoir; 7 – Discharge valves for purge; 8 – Discharge pump; 9 – Effluent outlet; 10 – Biogas outlet; 11 – Automation system; - - -Electrical connections]

Figure 1. Scheme of the AnSBBR with recirculation of the liquid phase (Lovato et al., 2017b).

2.2. Support for Immobilization and Inoculum

The inert support used consisted of 1-cm polyurethane foam cubes with an apparent density of 23 $kg.m^{-3}$ and a porosity of 95%. One of the main advantages of using polyurethane foam is that its high porosity allows immobilization of a significant amount of biomass that does not become detached during the charge, discharge, and reaction steps while the liquid phase is recirculated (Lovato et al., 2016).

The inoculum used in all experiments came from an up-flow anaerobic sludge blanket (UASB) reactor treating wastewater from a poultry slaughterhouse. The immobilization procedure consisted in crushing the sludge through a 0.5-mm mesh nylon sieve by completely immersing the foam with the obtained suspension followed by intense homogenization and 2-h of rest. Poorly adhered solids were washed off, and the medium was drained (Zaiat et al., 2001). This inoculum presented total volatile solid and total solid concentrations of 51 and 62 $g.L^{-1}$, respectively. (Lovato et al., 2016).

2.3. Stability and Performance Indicators

All stability and performance indicators used in the comparison are described in Lovato et al. (2016): organic matter removal efficiency for total (ε_T) and soluble (ε_S) samples; applied organic loading rate (OLR_A), removed organic loading rate (OLR_R), molar productivity (MPr), specific molar productivity (SMPr); molar biogas yield hydrogen or methane per applied load (Y_A) and molar hydrogen or methane yield per removed load (Y_R).

2.4. Wastewater

The analyzed papers studied the co-digestion of three agro-industrial wastes: whey, glycerin and vinasse (sugarcane stillage). Hence, the investigated influents studied were composed of different proportions of these three agro-industrial wastes (whey/glycerin, whey/vinasse and vinasse/glycerin). The influent composition was expressed as COD. The raw concentrations of the agro-industrial wastes were approximately: 70 $gCOD.L^{-1}$ for whey, 1520 $gCOD.L^{-1}$ for glycerin and 25 $gCOD.L^{-1}$ for vinasse. Table 1 shows the experimental conditions used.

Table 1. Experimental conditions from the analyzed papers

Reference	Assay	Influent Composition (% in COD basis)			Influent Concentration ($mgCOD.L^{-1}$)	Feed Strategy	OLR_A ($kgCOD.m^{-3}.d^{-1}$)
		Whey (%)	Glycerin (%)	Vinasse (%)			
(Lovato et al., 2016)	1	100	0	0	5000	FB	7.5
	2	75	25	0			
	3	50	50	0			
	4	25	75	0			
	5	0	100	0			
(Lovato et al., 2019a)	6	100	0	0	5000	B	6.4
	7	75	0	25			
	8	50	0	50			
	9	25	0	75			
	10	0	0	100			
(Lovato et al., 2019b)	11	0	100	0	5000	FB	5.0
	12	0	67	33			
	13	0	50	50			
	14	0	33	67			
	15	0	0	100			

B – Batch (feed time 5% of cycle length – 10 min); FB – Fed batch (feed time 50% of cycle length – 4 h).

2.5. Energy Estimation

The energy assessment was done as presented in the work of Volpini et al. (2018). The energy estimation of the reactors considered the best condition (in relation to biogas productivity, system stability and limiting co-substrate proportion) reported in each of the analyzed papers (Table 2). In the energy estimation a dilution volume was considered for maintaining the operational conditions of the laboratory-scale reactors (Albanez et al., 2016b). To estimate the volume of the industrial-scale reactor, the design parameter used was the removed organic loading rate (OLR_R). The energy estimation of the reactors and the energy produced on an industrial-scale was performed considering the industrial waste production data presented in Table 3.

Table 2. Laboratory-scale reactors parameters used for energy estimation in industrial-scale

Parameter	Lovato et al. (2016)	Lovato et al. (2019a)	Lovato et al. (2019b)
Influent concentration – C_{CT} ($g.L^{-1}$)	5.1	4.9	4.7
Cycle length – t_C (h)	8.0	8.0	8.0
Feeding time (h)	4.0	0.2	4.0
Number of cycles per day – N_i	3.0	3.0	3.0
Total volume of liquid medium – V_R (L)	3.0	3.5	3.0
Volume fed per cycle – V_F(L)	1.5	1.5	1.0
Residual volume – V_{RES} (L)	1.5	2.0	2.0
Volume of biomass + support – V_{B+S} (L)	2.0	2.0	2.0
Total volume – V_T (L)	5.0	5.5	5.0
Effluent concentration – C_{CF} ($g.L^{-1}$)	0.6	0.5	0.4
Organic load rate – OLR_R ($kgCOD.m^{-3}.d^{-1}$)	6.8	5.5	4.3
Molar productivity – MPr ($molCH_4.m^{-3}.d^{-1}$)	101.8	63.8	59.6
Molar yield per removed organic load – Y_R ($molCH_4.m^{-3}.d^{-1}$)	15.0	11.6	13.9

Table 3. Industrial waste production

Waste	Waste production ($L \cdot month^{-1}$)
Glycerin	1,027,000
Whey	6,900,000
Vinasse	230,000,000

Federal Government of Brazil, 2017

This estimation approach, using industrial data, was carried out only with the objective to demonstrate a preliminary application of the results obtained and to provide information for full-scale use of the investigated technological configuration (AnSBBR). It should be mentioned that detailed studies concerning operational aspects have to be considered to give a precise prediction of the scale-up system, such as in Hewitt and Nienow (2007), who reported reduction efficiency in large scale systems for aerobic batch reactors and discussed the importance of impeller type, rotor speed and agitation mode to ensure mass transfer and cell integrity. The dependence between mass transfer, impeller type and consumed power per volume ($kW.m^{-3}$) were also studied by Michelan et al. (2009) using a bench-scale anaerobic batch reactor with granular biomass (ASBR), and Farias de Novaes et al. (2010) using a pilot-scale anaerobic batch reactor with granular biomass (ASBR) and immobilized biomass (AnSBBR).

3. Results and Discussion

3.1. Process Stability and Performance

Table 4 shows the monitored variables in the analyzed papers. The systems achieved, at all conditions, high and steady organic matter removal in the form of COD, from 83 to 95% for filtered samples (soluble), which is important for stable anaerobic processes. The lowest value of organic matter removal efficiency expressed as COD was obtained in the study of Lovato et al. (2019a) (influent composed of vinasse and whey).

Removal efficiency is a parameter used to verify the stability and performance of the methanogenic reactor, because to produce methane and carbon dioxide organic matter (COD) is consumed and intermediate acids are generated/consumed. Carbohydrate and glycerin were also removed (Lovato et al., 2019b, 2016) with removal efficiencies exceeding 92% for whey and glycerin in all assays.

The co-digestion did not affect system behavior with respect to pH, bicarbonate alkalinity (BA) and total volatile acids (TVA). Minimum and maximum effluent pH were 7.4 and 8.3, respectively, which is a relatively small variation. It is worth mentioning that these pH values are in accordance with the ideal pH for methanogenic reactors (Lettinga and Haandel, 1993). Bicarbonate alkalinity was generated during the cycle as result of the metabolic reactions for biogas generation, as expected for any well-functioning digester. No volatile fatty acids accumulation was observed in any of the assays.

Table 4. Organic matter removal and process stability

Reference	Assay	Influent Concentration ($g.L^{-1}$)			Removal Efficiency (%)			Influent		Effluent		
		COD	Carbohydrate	Glycerin	COD	Carbohydrate	Glycerin	BA	TVA	pH	BA	TVA
(Lovato et al., 2016)	1	4.9	3.5	-	90	99.2	-	2.1	0.24	7.7	2.2	0.26
	2	5.1	2.4	0.4	89	99.6	98.2	1.9	0.13	7.4	1.8	0.32
	3	5.2	1.3	0.8	94	99.5	98.4	2.1	0.11	7.5	2.1	0.10
	4	5.2	0.7	1.2	94	98.9	99.9	2.0	0.10	7.4	2.0	0.13
	5	5.3	-	1.7	87	-	99.9	2.1	0.11	7.8	2.0	0.14
(Lovato et al., 2019a)	6	5.0	-	-	87	-	-	2.8	0.22	7.5	2.8	0.18
	7	5.1	-	-	83	-	-	1.1	0.27	7.6	1.6	0.25
	8	4.9	-	-	88	-	-	1.3	0.36	7.7	1.9	0.12
	9	4.8	-	-	86	-	-	0.5	0.54	7.6	1.4	0.15
	10	5.1	-	-	86	-	-	1.9	0.88	8.0	3.4	0.17
(Lovato et al., 2019b)	11	5.0	-	3.47	92	-	99.5	1.3	0.04	7.7	1.2	0.15
	12	5.6	0.23	2.3	95	92	99.6	1.13	0.41	7.7	1.6	0.07
	13	4.7	0.4	2.57	92	92	99.5	1.31	0.41	7.5	1.82	0.08
	14	4.7	0.7	1.4	93	95	99.5	1.1	0.7	7.7	1.6	0.08
	15	4.7	1.1	-	88	94.8	-	0.9	1.0	8.3	2.5	0.08

BA in $gCaCO_3.L^{-1}$; TVA in $gHAc.L^{-1}$.

Table 5. Performance indicators related to biogas production

Reference	Assay	Performance Indicators		Biogas composition
		MPr ($molCH_4.m^{-3}.d^{-1}$)	**MYRL ($molCH_4.kgCOD^{-1}$)**	**CH_4 (%)**
(Lovato et al., 2016)	1	93.7	14.0	73.0
	2	101.8	15.0	71.7
	3	94.2	12.7	76.4
	4	78.2	10.7	77.1
	5	78.7	11.4	79.1
(Lovato et al., 2019a)	6	76.3	12.7	72.2
	7	73.8	12.3	78.2
	8	63.9	11.6	77.1
	9	53.9	9.5	78.9
	10	51.1	9.1	82.3
(Lovato et al., 2019b)	11	63.1	14.0	79.4
	12	60.2	12.6	81.9
	13	59.6	13.6	83.0
	14	58.6	13.3	83.5
	15	57.8	13.7	83.2

Table 5 shows the values obtained for the performance indicators related to biogas production.

In relation to the co-digestion of whey and glycerin (Lovato et al., 2016) the molar productivity varied between 78.2 and 101.8 $molCH_4$ $m^{-3}day^{-1}$ and methane content varied between 71.7 and 79.1%; the maximum value was achieved in assay 2 (25% of whey and 75% of glycerin). This is an enhancement of 7.9% when compared to the productivity of whey alone and 29.4% when compared to the productivity of glycerin alone; therefore, the co-digestion of whey and glycerin was efficient in improving the reactor performance, most likely as a result of the nutrient balance that occurred between the substrates. Assay 2 also achieved the best yield per applied organic load (13.3 $molCH_4$ $kgCOD^{-1}$) (Lovato et al., 2016).

Lovato et al. (2019a) observed a clear trend of methane molar productivity and yield decrease around 30% when vinasse is added to the influent (Assays 6 to 10, 100% whey and 100% vinasse, respectively). In spite of this, vinasse improved the methane content in biogas. These results show the difficulty in producing biogas from the anaerobic digestion of vinasse and the need for co-digestion with a more biodegradable co-substrate like whey. This same behavior was observed in the work of Albanez et al. (2016b) in which the objective was to produce hydrogen from the anaerobic treatment (in an AnSBBR) of a vinasse and molasses based wastewater. The negative effect of vinasse can be credited to the low biodegradability of this effluent and the presence of recalcitrant compounds.

According to Lovato et al. (2019b) methane productivity and yield decreased linearly as more vinasse was added to the influent, as observed by Lovato et al. (2019a). Assays 11 (100% glycerin) and 15 (100% vinasse) were carried out to verify how the mono-digestion of these wastewaters would perform regarding methane production. Methane productivity in assay 11 (63.08 mol CH4 m^{-3} day^{-1}) was higher (around 9%) than in assay 15 (57.83 mol CH4 m^{-3} day^{-1}). The increase in methane productivity and yield comparing glycerin and vinasse mono-digestion is credited to the presence of recalcitrant organic compounds in the latter. Thus, glycerin proved to be a good co-substrate for vinasse since it increased the overall methane production in assays 12, 13, and 14, probably due to the increase in degradable organic loading rate. The authors suggested that a good work ratio of glycerin and vinasse would be 50:50, since assay 13 achieved an intermediate methane productivity and yield when compared to both mono-digestions, and this proportion represents only 1.53 v% glycerin in the influent, rendering glycerin storage inside the vinasse treatment plant feasible.

Hence, using glycerin as a co-substrate for whey and vinasse improved reactor performance, regarding methane productivity and yield. An optimum percentage of whey and glycerin could be determined (75% whey and 25% glycerin) (Lovato et al., 2016). On the other hand, in the case of vinasse as the main substrate, an optimum ratio between vinasse and glycerin could not be defined. Since vinasse is difficult to treat, glycerin addition improved reactor performance (up to a glycerin content of 100%). Therefore, Lovato et al. (2019b) proposed as an acceptable composition in the influent a proportion of 50% vinasse and 50% glycerin.

Vinasse showed to be the substrate (among whey, glycerin and vinasse) with the worst performance results for the investigated systems. Therefore, vinasse is not recommended as co-substrate for either whey or glycerin. It is important to note that a considerable improvement in reactor performance was observed when whey or glycerin was used as co-substrate in the treatment of vinasse. Comparing assays 8 (50% vinasse and 50% whey) and 13 (50% vinasse and 50% glycerin) showed that whey as co-substrate resulted in higher molar productivity values than when glycerin was used as co-substrate. On the other hand, the yield was higher when glycerin was used as a co-substrate.

3.2. Energy Estimation

Table 6 shows the volumes of the laboratory-scale and estimated industrial-scale methanogenic reactors for the investigated papers.

Table 6. Results of the estimation of industrial-scale reactors

Reference	**Lovato et al. (2016)**		**Lovato et al. (2019a)**		**Lovato et al. (2019b)**	
Scale	Laboratory	Industrial	Laboratory	Industrial	Laboratory	Industrial
C_{CT} (gCOD.L^{-1})	5.1	91.9	4.9	35.1	4.7	46.1
t_C (h)	8	8	8	8	8	8
OLR_R (kgCOD.m^{-3}.d^{-1})	6.8	6.8	5.7	5.7	4.3	4.3
V_F (m^3.cycle^{-1})	$1.5 \cdot 10^{-3}$	68	$1.5 \cdot 10^{-3}$	3,448	$1.0 \cdot 10^{-3}$	2,624
V_{RES} (m^3)	$1.5 \cdot 10^{-3}$	2,680	$2.0 \cdot 10^{-3}$	59,770	$2.0 \cdot 10^{-3}$	81,093
V_R (m^3)	$3.0 \cdot 10^{-3}$	2,748	$3.5 \cdot 10^{-3}$	6,3218	$3.0 \cdot 10^{-3}$	83,717
V_{B+S} (m^3)	$2.0 \cdot 10^{-3}$	1,832	$2.0 \cdot 10^{-3}$	36,125	$2.0 \cdot 10^{-3}$	55,812
V_T (m^3)	$5.0 \cdot 10^{-3}$	4,580	$5.5 \cdot 10^{-3}$	99,343	$5.0 \cdot 10^{-3}$	139,529

Lovato et al. (2016) proposed to treat all the whey produced by a dairy industry with a productivity of 6000 m³ of milk per month (large industry). To implement co-digestion, addition to the wastewater of 92.1 m³ of glycerin per month would be required to reach the 75:25 (COD basis) proportion. This amount of glycerin can be easily stored inside the plant, as it has physical/chemical properties suitable for storage (dense, nonexplosive liquid) and currently biodiesel producers have to pay to discard all the exceeding glycerin from their plants, because of the large quantities produced and its difficulty in being sold. Four methane-producing AnSBBRs with the same volume are proposed (4,580 m^3, i.e., 4 reactors of 1,145 m^3 each – Table 6) operating in parallel to adapt the discontinuous operation into a continuous operation system. In this configuration, a reactor would always being fed, another being discharged and the last two would promote biotransformations, completing three steps: feeding, reaction and discharge. Cycle length was maintained at 8 h: 2.0 h for feeding, 2.0 h for discharge and 4.0 h for the bioreaction alone. Such a configuration would allow continuous biogas production.

Lovato et al. (2019b, 2019a) proposed to treat all the vinasse produced by a sugar and alcohol industry with a productivity of 230,000 m^3 of vinasse per month (mid-size industry). In this case, co-digestion would require addition of 78,800 m^3 of whey per month (Lovato et al. 2019a) or 3,600 m^3 of glycerin (Lovato et al. 2019b) to reach the 50:50 (COD basis) proportion.

Co-digestion of vinasse with whey would require four methane-producing AnSBBRs (99,343 m^3, i.e., 4 reactors of 24,836 m^3 each) operating in parallel to adapt the discontinuous operation into a continuous operation system. Regarding the co-digestion of vinasse with glycerin, four AnSBBRs with the same volume (139,529 m^3, i.e., 4 reactors of 34,882 m^3 each) and operation scheme would be required.

The energy yields achieved by the systems would be 12.0 $MJ.kgCOD_{removed}^{-1}$ (Whey + Glycerin), 9.1 $MJ.kgCOD_{removed}^{-1}$ (Vinasse + Whey) and 11.1 $MJ.kgCOD_{removed}^{-1}$ (Vinasse + Glycerin), considering the four reactors working in parallel from the scale-up estimation.

The biogas generation would be 277.4 $kmolCH_4.d^{-1}$, 4,143.8 $kmolCH_4.d^{-1}$ and 4,989.4 $kmolCH_4.d^{-1}$, respectively, therefore proving that glycerin is in fact a better co-substrate for vinasse than whey. If the biogas from the co-digestion of vinasse and whey were applied for electricity generation in stationary engines (38% efficiency (Moraes et al., 2015)), the energy generated would reach approximately 107,000 MW·h per year, which is enough electricity to supply the energy demand of a city with 41,000 inhabitants.

CONCLUSION

An overview was performed of achievements of three papers that used AnSBBRs to co-digest the agro-industrial wastes vinasse, whey and glycerin for methane production, focusing on operational strategy and perspectives for energy estimation.

Regarding the co-digestion of whey and glycerin (Lovato et al., 2016), the molar productivity varied between 78.2 and 101.8 $molCH_4.m^{-3}.d^{-1}$ and methane content varied between 71.7 and 79.1%; the maximum value was achieved at a proportion of 25% whey and 75% glycerin. This is an enhancement of 7.9% when compared to the productivity of whey alone and 29.4% when compared to the productivity of glycerin alone.

When whey was used as a co-substrate for vinasse (Lovato et al., 2019a), a clear trend of decreasing methane molar productivity and yield was observed when vinasse was added to the influent. This effect was also observed in the work of Lovato et al. (2019b), where glycerin was used as a co-substrate for sugarcane stillage. The intermediate (50:50 ratio) methane productivity was 63.9 $molCH_4$ $m^{-3}day^{-1}$ for the system treating whey and vinasse and 59.6 $molCH_4$ $m^{-3}day^{-1}$ for the system containing vinasse and glycerin.

The energy yields achieved by the industrial-scale systems would equal 12.0 $MJ.kg.COD_{removed}^{-1}$ (Whey + Glycerin), 9.1 $MJ.kgCOD_{removed}^{-1}$ (Vinasse + Whey) and 11.1 $MJ.kgCOD_{removed}^{-1}$ (Vinasse + Glycerin), considering four reactors working in parallel from the scale-up estimation.

The biogas generation would be 277.4 $kmolCH_4.d^{-1}$, 4,143.8 $kmolCH_4.d^{-1}$ and 4,989.4 $kmolCH_4.d^{-1}$, respectively, therefore proving that glycerin is in fact a better co-substrate for vinasse compared to whey. If the biogas from the co-digestion of vinasse and whey were applied for electricity generation in stationary engines, the energy generated would reach approximately 107,000 MW·h per year, which is enough electricity to supply the energy demand of a city with 41,000 inhabitants.

Acknowledgments

This study was supported by the São Paulo Research Foundation (FAPESP #15/06.246-7), the National Council for Scientific and Technological Development (CNPq #443181/2016-0) and the Coordination for the Improvement of Higher Education Personnel (CAPES).

References

Albanez, R., Chiaranda, B.C., Ferreira, R.G., França, A.L.P., Honório, C.D., Rodrigues, J.A.D., Ratusznei, S.M., Zaiat, M., 2016a. Anaerobic Biological Treatment of Vinasse for Environmental Compliance and Methane Production. *Appl. Biochem. Biotechnol.* 178, 21–43. doi:10.1007/s12010-015-1856-z.

Albanez, R., Lovato, G., Zaiat, M., Ratusznei, S.M., Rodrigues, J.A.D., 2016b. Optimization, metabolic pathways modeling and scale-up estimative of an AnSBBR applied to biohydrogen production by co-digestion of vinasse and molasses. *Int. J. Hydrogen Energy* 41, 20473–20484. doi:10.1016/j.ijhydene.2016.08.145.

Carvalho, F., Prazeres, A.R., Rivas, J., 2013. Cheese whey wastewater: Characterization and treatment. *Sci. Total Environ.* doi:10.1016/j.scitotenv.2012.12.038.

Chandra, R., Takeuchi, H., Hasegawa, T., 2012. Methane production from lignocellulosic agricultural crop wastes: A review in context to second generation of biofuel production. *Renew. Sustain. Energy Rev.* doi:10.1016/j.rser.2011.11.035.

Farias de Novaes, L., Saratt, B.L., Rodrigues, J.A.D., Ratusznei, S.M., de Moraes, D., Ribeiro, R., Zaiat, M., Foresti, E., 2010. Effect of impeller type and agitation on the performance of pilot scale ASBR and AnSBBR applied to sanitary wastewater treatment. *J. Environ. Manage.* doi:10.1016/j.jenvman.2010.03.008.

Federal Government of Brazil, 2017. Paraná Institute for Economic and Social Development 2017 [WWW Document]. URL http://www.ipardes.gov.br/ (accessed 7.20.12).

Ferraz Junior, A.D.N., Wenzel, J., Etchebehere, C., Zaiat, M., 2014. Effect of organic loading rate on hydrogen production from sugarcane vinasse in thermophilic acidogenic packed bed reactors. *Int. J. Hydrogen Energy* 39, 16852–16862. doi:10.1016/j.ijhydene.2014.08.017.

Hewitt, C.J., Nienow, A.W., 2007. The Scale-Up of Microbial Batch and Fed-Batch Fermentation Processes. *Adv. Appl. Microbiol.* doi:10.1016/S0065-2164(07)62005-X.

Leite, J.A.C., Fernandes, B.S., Pozzi, E., Barboza, M., Zaiat, M., 2008. Application of an anaerobic packed-bed bioreactor for the production of hydrogen and organic acids. *Int. J. Hydrogen Energy.* doi:10.1016/j.ijhydene.2007.10.009.

Lettinga, G., Haandel, A.C. van, 1993. Anaerobic digestion for energy production and environmental protection. *Renew. energy sources fuels Electr.* 817–839..

Lovato, G., Albanez, R., Albuquerque, J.N., Cola, P., Celestino, R.S., Vogel, S.E., Fukuyama, M., Hirata, F.E., Saito, F.H., Ratuznei, S.M., Rodrigues, J.A.D., 2017a. Novel Insights into the Co-Digestion of Whey with Glycerin in an AnSBBR: Influent Composition and Concentration, Cycle Length and Feed Strategy Effect Title. In: Daniels, J.A. (Ed.), *Advances in Environmental Research* - Volume 58. Nova Science Publishers, Hauppauge, pp. 161–182.

Lovato, G., Albanez, R., Lima, D.M.F., Bravo, I.S.M., Almeida, W.A., Ratuznei, S.M., Rodrigues, J.A.D., 2015. Application and environmental compliance of anaerobic sequencing batch reactors applied to hydrogen/methane bioenergy production, 1st ed, *Wastewater*

Treatment: Processes, Management Strategies and Environmental/ Health Impacts. Nova Science Publishers, Hauppauge.

Lovato, G., Albanez, R., Triveloni, M., Ratusznei, S.M., Rodrigues, J.A.D., 2019a. Methane Production by Co-Digesting Vinasse and Whey in an AnSBBR: Effect of Mixture Ratio and Feed Strategy. *Appl. Biochem. Biotechnol.* 187, 28–46. doi:10.1007/s12010-018-2802-7.

Lovato, G., Batista, L.P.P., Preite, M.B., Yamashiro, J.N., Becker, A.L.S., Vidal, M.F.G., Pezini, N., Albanez, R., Ratusznei, S.M., Rodrigues, J.A.D., 2019b. Viability of Using Glycerin as a Co-substrate in Anaerobic Digestion of Sugarcane Stillage (Vinasse): Effect of Diversified Operational Strategies. *Appl. Biochem. Biotechnol.* doi: 10.1007/s12010-019-02950-1.

Lovato, G., Lazaro, C.Z., Zaiat, M., Ratusznei, S.M., Rodrigues, J.A.D., 2017b. Biohydrogen production by co-digesting whey and glycerin in an AnSBBR: Performance optimization, metabolic pathway kinetic modeling and phylogenetic characterization. *Biochem. Eng. J.* 128, 93–105. doi:10.1016/j.bej.2017.09.011.

Lovato, G., Ratusznei, S.M., Rodrigues, J.A.D., Zaiat, M., 2016. Co-digestion of Whey with Glycerin in an AnSBBR for Biomethane Production. *Appl. Biochem. Biotechnol.* 178, 126–143. doi:10.1007/ s12010-015-1863-0.

Mata-Alvarez, J., Dosta, J., Romero-Güiza, M.S., Fonoll, X., Peces, M., Astals, S., 2014. A critical review on anaerobic co-digestion achievements between 2010 and 2013. *Renew. Sustain. Energy Rev.* 36, 412–427. doi:10.1016/j.rser.2014.04.039.

Merlin, G., Boileau, H., 2013. Anaerobic Digestion of Agricultural Waste: State of the Art and Future Trends, in: Torrles, A. (Ed.), *Anaerobic Digestion: Types, Processes and Environmental Impact.* Nova Science Publishers.

Michelan, R., Zimmer, T.R., Rodrigues, J.A.D., Ratusznei, S.M., de Moraes, D., Zaiat, M., Foresti, E., 2009. Effect of impeller type and mechanical agitation on the mass transfer and power consumption aspects of ASBR operation treating synthetic wastewater. *J. Environ. Manage.* 90, 1357–1364. doi:10.1016/j.jenvman.2008.08.003.

Moraes, B.S., Zaiat, M., Bonomi, A., 2015. Anaerobic digestion of vinasse from sugarcane ethanol production in Brazil: Challenges and perspectives. *Renew. Sustain. Energy Rev.* 44, 888–903. doi:10.1016/j.rser.2015.01.023.

Panesar, P.S., Kennedy, J.F., Gandhi, D.N., Bunko, K., 2007. Bioutilisation of whey for lactic acid production. *Food Chem.* doi:10.1016/j.foodchem.2007.03.035.

Rico, C., Muñoz, N., Fernández, J., Rico, J.L., 2015. High-load anaerobic co-digestion of cheese whey and liquid fraction of dairy manure in a one-stage UASB process: Limits in co-substrates ratio and organic loading rate. *Chem. Eng. J.* 262, 794–802. doi:10.1016/j.cej.2014.10.050.

Rivero, M., Solera, R., Perez, M., 2014. Anaerobic mesophilic co-digestion of sewage sludge with glycerol: Enhanced biohydrogen production. *Int. J. Hydrogen Energy* 39, 2481–2488. doi:10.1016/j.ijhydene.2013.12.006.

Shah, F.A., Mahmood, Q., Rashid, N., Pervez, A., Raja, I.A., Shah, M.M., 2015. Co-digestion, pretreatment and digester design for enhanced methanogenesis. *Renew. Sustain. Energy Rev.* doi:10.1016/j.rser.2014.10.053.

Volpini, V., Lovato, G., Albanez, R., Ratusznei, S.M., Rodrigues, J.A.D., 2018. Biomethane generation in an AnSBBR treating effluent from the biohydrogen production from vinasse: optimization, metabolic pathways modeling and scale-up estimation. *Renew. Energy* 116 (Pt A), 288–198. doi:10.1016/j.renene.2017.09.004.

Zaiat, M., Rodrigues, J.A.D., Ratusznei, S.M., de Camargo, E.F.M., Borzani, W., 2001. Anaerobic sequencing batch reactors for wastewater treatment: a developing technology. *Appl. Microbiol. Biotechnol.* 55, 29–35. doi:10.1007/s002530000475.

In: Wastewater Treatment
Editor: Adriana Magdalena
ISBN: 978-1-53616-370-4

Chapter 7

RECLAMATION AND REUSE OF GREY WATER: A BENEFICIARY SOLUTION FOR THE WATER DEMAND ISSUES

Rajagopalan Varadarajan[1,*], Manjula Gopinathan[2] and Abirami Mani[1]

[1]Department of Civil Engineering, Anna University BIT Campus, Tiruchirappalli, TN, India

[2]Department of Civil Engineering, SVS College of Engineering, Coimbatore, TN, India

ABSTRACT

As per the world water development report the global demand for water has been increasing at rate of 1% for every year due to population growth, economic development and changing consumption pattern. Increasing the demand of water due to exponential growth of population led to the idea to recycle of waste water. In Wastewater treatment greywater recycling is rising as an integral part of water demand

* Corresponding Author's E-mail: vrgopalanaut@gmail.com.

management, contributing to the progress of the preservation of high quality fresh water and also reducing pollutants in the environment and lowering the overall supply costs. An alternative source for treasured potable water is grey water. Grey water is the waste water from homes excluding black water. Typically 50 to 80% of waste water is grey water. This chapter will discuss about the character of grey water and available methods to recycle and reuse grey water.

Keywords: grey water, microfiltration, physico-chemical treatment, phytotoxicity, reuse

1. INTRODUCTION

Freshwater supply is essential to sustain economic activities and the well-being of humankind. Nowadays water quality and scarcity have become major concerns in most countries, in particular developing countries. Of the total available water on the earth, approximately 97% is salty, that is not suitable for human consumption. Of the 3% of freshwater, only one third confirms to drinking water quality standards which can sustain human's daily lives and for other usage (Ding 2017). Wastewater reclamation and reuse techniques have been applied in many countries because of low rainfall and high evaporation due to high demand of water (Al-Zouby et al. (2017)). According to CPHEO manual 1993, for an average Indian household, water requirement is 135 lpcd (Srivastava (2018)). Every day, grey water of about 70 to 90 litres will be generated as waste water. With the increasing demand, water has become an important resource in the world. According to International water management institute one person in three will live in water scarcity condition by 2025 (Deshmukh et al. (2017)). Grey water reuse is one of the sustainable ways to reduce the usage of potable water and the reclamation of grey water is an important aspect in environmental engineering. Moreover use of treated grey water in the field of agriculture can reduce the consumption of fresh water (Deshmukh et al. (2017)). Grey water is waste water which is different from black water. The main difference between grey water and black water is, grey water contains only about a tenth of the nitrogen, it contains low load of fecal micro organisms (Laaffat et al. (2019)). Treatment technologies include different systems such as pretreatment and post treatment, whereas the disinfection process is used to reduce the microbiological parameters (Fouad et al. (2017)). After treatment, UV disinfection method was employed in pilot-scale municipal wastewater treatment for subsequent agricultural applications and removing the health risk (Bhattacharya et al. (2013)).

2. Grey Water Characteristics

Grey water is the waste water from household excluding black water and it does not contain any fecal matter. It constitutes about 60-70% of domestic waste water (Deshmukh et al. (2017)). Grey water reclamation and reuse is a sustainable solution that could be utilized for non-potable uses such as watering the garden, playing fields and lawns, water closet flushing, recharging fountains and also for outdoor washing (James and Ifelebuegu (2018)). Municipal wastewater is generally divided into yellow water, brown water and grey water. Among them, yellow and brown water refer to urine and fecal sewage. Grey water includes streams from showers/baths, wash basins, laundry, kitchen sinks and dishwashers and is generally defined as urban wastewater without any pollution from toilets (Kuang-Wei et al. (2018)). The composition of grey water depend upon resources from where the water drawn and also the type of household activities. It has chemicals and several millions of pathogenic bacteria which can cause a health hazard if this water is reused without proper treatment. Grey water contains phosphorus and nitrates due to the usage of detergent, shampoo and dishwasher liquid, bacteria, total suspended solids, oil and grease, total suspended solids (TSS), hair from the bathrooms, dirt from laundry and food particles from kitchen waste (Arugam et al. (2018)).

However, grey water recycle have some benefits. It poses some risks and challenges that must be mitigated for safe reuse. Mostly, greywater contains high concentrations of biodegradable organic contents and nutrients from households like nitrates, phosphorus and xenobiotic organic compounds, biological microbes, salmonella and general hydro chemical constituents. Due to the modern lifestyle of people it may also contain wastes like aerosols, pigments, beauty products and also toxic heavy metals (Oteng-Peprah et al. (2018)). Grey water charateristics from different sources are given in Table 2.1. While treating grey water, important issues like presence of pathogenic microorganisms and environmental risks from the presence of pollutants such as sodium, surfactants should be addressed (Maimon and Gross (2018)). The materials such as charcoal, coconut shell, feldspar, kaolinite, alumina, betonies, pulverized fuel ash, elite, and tire granules are promising as potential sustainable materials for the reclamation of grey water (James and Ifelebuegu (2018)).

Table 1. Grey water characteristics

Source	BOD	COD	Turbidity	NH_3	Total coliforms
Hand basin	109	263	-	9.6	-
Combined	121	371	69	1	-
Synthetic greywater	181	-	25	0.9	$1.5x10^6$
Single person	110	256	14	-	-
Single family'	-	-	76.5	0.74	-
Flats	33	40	20	10	$1x10^6$
College	80	146	59	10	-

Source: Al-Jayyousi (2003).

3. Methods for Reclamation of Grey Water

Grey water samples are collected from the hostel sources. The samples were analyzed for physicochemical and microbial parameters. The parameters pH and conductivity were all measured in-situ. All samples were stored in an ice chest packed with ice cubes below 4 ^{0}C and transported to the laboratory within 4 hours. Water quality standards such as Total Suspended Solids (TSS), Chemical Oxygen Demand (COD), Turbidity and Biochemical Oxygen Demand (BOD_5) Ca, Mg, Na, Heavy metals, Organic pollutants, Nitrogen, Phosphate were tested in laboratory (Dwumfour-Asare et al. 2017).

In comparison with the traditional centralized wastewater treatment, the decentralized process is better in both operating and initial cost because no large sewage pipe system is needed. The decentralized techniques are generally adopted for grey water reclamation process. It can be categorized into three groups:

- Physical treatment
- Chemical treatment
- Biological treatment

Physical treatments are coarse sand filtration and membrane filtration. Chemical treatments are coagulation, photo-catalytic oxidation, and ion exchange and then finally biological treatments are Constructed Wetlands (CW), Membrane Bioreactors (MBR), Rotary Biological Contactors (RBC) and Sequencing Batch Reactors (SBR) (Zhu et al. (2018)). MBR appears to be an exact and feasible solution in urban residential buildings (Li et al. (2009)). Physical screens with large-pore filters and simple devices such as metal screens and nylon socks used to remove large particles from Greywater (Zhu et al. (2018)). Micro-ultra and nano-filtration and reverse osmosis (RO) are Pressure-driven membrane processes.

Physical processes alone are not sufficient for reduction of the organics and nutrients in the greywater; Chemical processes also should be included with it to attain efficient removal of suspended solids;physical filtration with aerobic system and disinfection treatment is the most feasible solution for grey water recycling.. Contribution of microbial pathogens and nutrient discharge into the environment from insufficient on-site wastewater treatment systems has raised concern in many areas due to the pollution of the nearby water recipient (Moges et al. (2017)).

Finally disinfection processes are carried. Ultraviolet irradiation is a physical disinfection process that achieves disinfection by inducing photo biochemical changes within microorganisms. It's a transfer of electromagnetic energy from a UV lamp to organism's genetic material (DNA and RNA). The energy absorbed generate photoproducts such as thymine dimmers on the same nucleic acid strand which blocked DNA replication and leading to inactivation of microorganism. Another widely used disinfection method is chlorine disinfectant which is easily available. It is useful to prevent the transmission of disease causing micro organism. Efficiency of Grey water treatment is examined by total coli form inactivation. Efficiency depends upon the particle size. It can remove about 91% total coliform. Free residual chlorine is also maintained for the prevention of future contamination. Disinfectant demand is affected by the organic concentration but it does not influence the microbial resistance to inactivation (Winwarda et al. (2008)). Reclaimed grey water which is used for toilet flushing should be treated by filtration and disinfection process. According to EPA, fecal coliforms in 100 ml of the treated water is should be nil and BOD should be less than 10 mg/l and residual chlorine should be greater than 1 mg/l whereby Cl should be continuously monitored. After the treatment, water should be used immediately before anaerobic conditions develop (Oteng-Peprah et al. (2018)).

4. Impact of Untreated Grey Water

Grey water should be treated properly, otherwise it causes major health risk and pollution of environment. Grey water may be harmful to some crops but suitable for others. Use of irrigation water of unsuitable quality will reduce crop yield and worsen the physical properties of the soil. Water with high solids causes clogging to the drip emitters, similarly high nutrient level enhances growth of bacteria inside the irrigation pipes and might clog it (Laaffat et al. (2019)). Grey water contains high value of SAR (sodium absorption ratio) and salinity which can deteriorate the soil structure and decrease the soil permeability. It also reduces the crop yields.

In long term, effects include accumulation of surfactants and salts which will affect the plants growth and soil micro organisms (Al-Hamaiedeh and Bino (2010)). Untreated greywater have impacts on plant biomass and nutrition, soil enzyme activity and worm avoidance. While salt content is accumulating in soil it indirectly elevates the concentration of iron and copper. Fe and Cu consumption by the plants lead to risk to human health and plant life. While considering pH of grey water, higher pH can affect soil pH and micronutrients in soil. Acceptable pH level of soil is 4-9 (Reichman and Wightwick (2013)).

5. Previous Case Studies in Grey Water Reclamation

5.1.Physical Treatment

5.1.1. Design of Packaged Grey Water Treatment Using Slow Sand Filter

It consists of designing, 3D modeling, prototyping and experimental evaluation. The designing involves the designing of pre-treatment stage with wire mesh, slow sand filtration and pump. The design is modeled into 3D prototype using Solid Works Ver. 2016. The performance of the design is evaluated using the prototype. The pre-treatment filter consists of coarse and fine wire mesh. Coarse wire mesh is 1 mm in pore size and the fine wire mesh pore size is 0.5 mm. Slow sand filter consists of sand and gravel bed which is used convert the organic load into biomass and gas which is trapped in the sand particles. Here slow sand filter design is based on the volumetric flow rate equation,

Volumetric flow rate indicates the volume of water passing through medium per unit time. The volumetric flow rate equation (1) is shown below.

$$A = \frac{Q}{v} \quad (1)$$

Where,

A = sand filter surface area (m^2)

Q = flow rate onto the filter surface (L/h)

v = loading rate of sand (L/hm^2)

Darcy's law is used to calculate the flow rate of liquid that passes through a porous medium as shown in equation (2).

$$A_{max} = ka.\frac{h+d}{D} \tag{2}$$

Where,

A_{max} = Maximum flow rate (m^3/s)

k = Hydraulic conductivity of sand (m/s)

a = Cross sectional area of sand filter (m^2)

h = Depth of clearance above the sand (m)

D = Overall depth of filter (m)

d = Depth of sand (m)

And then third parameter pump design is based on water level in collecting tank.

Otherwise pump design is based on head loss and horse power. That can be calculated by using equation (3).

$$WHP = Q.\frac{H}{3960} \tag{3}$$

Where,

WHP = Water horsepower (hp)

Q = flow rate of water from the pump (gallon/minute)

H = head loss or distance to pump the water (foot)

The material budget of a complete treatment system is lower than the existing home scale greywater recycling system. This design is able to treat greywater to meet the standard water quality of toilet flushing and also irrigation purpose (Deshmukh et al. (2017)).

5.1.2. Membrane Process

Membrane-based processes are being widely used for grey water treatment. Submerged hollow fiber membrane module had been extensively studied for the treatment of domestic wastewater (Zhu et al. (2018)). There are four processes currently used namely, ultra filtration (UF), reverse osmosis (RO), nanofiltration (NF), and microfiltration (MF)

5.1.2.1. Microfiltration

In this study feed flow rate on permeate flux and rejection characteristics have been studied. Wastewater was treated through a membrane process using a mixed cellulose ester (MCE) microfiltration membrane with pore size of 0.22 μm. The TMP (transmembrane pressure) decreases with an increase in feed flow rate. The BOD, COD, TSS and turbidity removal efficiency rate was obtained about 93.9, 90.8 and 98.7 percent, respectively which was obtained at a TMP of 1 bar and feed flow rate of 44 L/h modeling. The structure of the membrane is symmetric and sponge-like with nearly uniform pores (Manouchehri and Kargari (2017))

5.2. Chemical Treatment

Chemical treatment for grey water can be done by coagulation and oxidation. One of the feasible solutions for grey water treatment is an advanced oxidation process based on photo-catalytic oxidation with suitable photo catalyst (titanium dioxide and ZnO) in the presence of UV light or visible light, was applied for grey water treatment resulting in about 90% removal of the organic matter (Chen et al. (2017)).

5.2.1. Coagulation

Coagulation and ion exchange methods are useful for treating the low strength greywater. Alum and ferric chloride are the major coagulants used for treatment. Coagulation mechanism contains two processes, one is charge neutrilisation for colloidal material at optimum pH and second one is charge complexation for soluble material. In both cases preferentially charged materials are removed. It can remove about 85% BOD. and about 74% COD. Complexation reaction between organic compounds and metal are known to be fast rather than hydroxide. The optimum level of alum and ferric chloride for shower water is 7 and 3.9 respectively (Pidou et al., (2008)).

5.2.2. Treatment of Grey Water by Using Electro Synthesized Ferrate (VI) Ion

Advanced oxidation is the effective method to treat grey water but due to expensive, it is not widely practiced. So electro coagulation method was used due to its removal rate. It removes about 70% COD. Ferrate ion (VI) has many advantages due to its dual function as oxidant and also a coagulant. Ferrate (VI) ion have reduction potential of 2.20 v in acidic condition. When Ferrate (VI) ion is oxidised, it is reduced to Fe (III) which is a well known coagulant in waste water treatment. Ferrate (VI) production method involves three methods such as wet chemical synthesis, dry chemical synthesis and electrochemical synthesis. Among the three methods, electrochemical method is mostly adapted due to its simplicity and absence of hypochlorites. It does not require any expensive chemicals. Ferrate (VI) was generated by electrochemical method in rectangular Plexiglas reactor and it is used for treatment simultaneously. NAOH is an alkaline media, and it has active surface area about 81.25 cm^2, and applied current density was 1.47 am cm^{-2}. Then 1 L of the greywater to be treated was placed in the jar tester. Ferrate (VI) was dosed into the jar. Next, the pH was adjusted to 2 and some buffers also used. Flocculation of samples was carried out for about 20 mins at 30 rpm and then allowed to settle for additional 60 minutes. To assess the performance of grey water treatment by ferrate (VI) in terms of percentage improvement of water quality and removal of organics and physical impurities, parameters such as COD, TOC and turbidity were considered. The experiments were conducted at pH 7 and temperature 25°C. TOC was removed in the range of 21.6–61.7% and turbidity removal was between 97 and 99.9%. The highest COD removal efficiencies were obtained at pH 7 and 8 as 88.8 and 87.6%, respectively.

In the irrigation field application, water containing surfactants may produce negative effects. It is recommended that the concentration of anionic surfactants should not exceed 1 mg/L. According to the results; Fe (VI) at the 75 mg/L dose is capable of leaving only 0.33 mg/L of surfactant in treated greywater. The optimum ferrate (VI) dose was found to be 75 mg/L, and the optimum pH was 7 greywater treatments (Barısci et al. (2015)).

5.2.3. Greywater Recycling with Al (III) Salt Combination with Fe (VI)

Fe (VI) is the efficient method in treat the grey water but due to the instability of ferrate solution and high production cost of solid ferrate products. the combination use of Fe (VI) and Al (III) salts was proved to be more efficient than using the Fe (VI) salts alone at greywater recycling. The optimum dosage of Fe (VI)/Al (III) salts was 25 mg/L for light greywater, 90/60 mg/L for dark greywater, respectively (Song et al. (2017)).

5.2.4. Treatment of Grey Water by Photo Catalyst Titanium Dioxide

Photo catalytic oxidation is a technology that could be suitable to remove organic compounds found in greywater. Currently titanium dioxide photo catalyst shows wide attention for photo catalytic oxidation of organic matter of wastewater because it is biologically and chemical inert, resistant to chemical corrosion and can work at ambient temperature and pressure, without addition of chemical. The irradiation of titanium dioxide dispersions by ultraviolet (UV) (300-400 nm) light can lead to the formulation of highly reactive hydroxyl radicals which attack the pollutant molecule to degrade it into carbon dioxide, water and mineral acids (Bhadiyadra and Vaghani (2015)).

5.2.5. Hybrid Electrochemical with GAC

Electrochemical reactor filled with granular activated carbon (GAC) was employed for the treatment of of GAC by the electric field, without the direct contact with the terminal current feeders, resulted in oxidant production and degradation of the adsorbed/electro absorbed contaminants

GAC was reduced from 72% to 22% when an electric field was applied; further confirming that electrochemical polarization of GAC caused the killing of the adsorbed microorganisms. Here Simultaneous adsorption/electro sorption and electro catalysis in the GAC packed bed are carried (Andres et al. (2018)).

5.3. Biological Treatment

5.3.1. Constructed Wetland

Constructed wetland is an artificially created wetland which is utilizing flora and fauna for treatment, utilizing ecological technology to mimic conditions. Constructed wetlands (CW) are also one of such systems considered as sustainable, cost effective greywater treatment for small communities. There are various types of constructed wetland classified based on their flow pattern; I) Horizontal subsurface flow constructed wetland, ii) Vertical subsurface flow constructed wetland and iii) Hybrid subsurface flow constructed wetland. Hybrid flow CW is mostly used. Waste water first flows into horizontal flow and then vertical flow. The advantage of the hybrid system is that the nitrogen can be nitrified completely in vertical flow CW and denitrified in horizontal flow CW (Ramprasada et al. (2017)). Biological/Physical Constructed Wetlands (CW) is a system in which GW is passed by gravitational flow through planted granular media on which microorganisms grow and degrade organic matter (Zhu et al. (2018)). Compared to conventional wastewater treatment technologies, constructed wetlands are mechanically simple and have relatively low operation and maintenance (O&M) requirements (Laaffat et al. (2019)). In that way vertical flow constructed wetland showed an overall removal efficiency rate of 81% for COD and 70% for TN and TP, and the treated GW sufficiently to meet current standards for unlimited irrigation (Chen et al. (2017)).

In this case study horizontal flow constructed wetland has been used to the grey water and up disinfection method was used. The horizontal sub-surface flow Constructed wetland (HSSFCW) reactor has been used to treat 1.2 m^3/d of GW for 100 days. In this study three lawn plots are irrigated with three various type water such as tap water, raw grey water and treated grey water and then risk analysis were performed to find the variation in growth of plants. The treated grey water with up disinfection shows significant growth. Concerning the disinfection, the experimental results obtained in this study showed that 50 maws/cm^2 UV dose is more enough to meet the microbial quality standards for water irrigation. However, constructed wetland removes low rate of Na, Ca, Mg and it also leads to increasing electrical conductivity by dissolution of organic matter in treated water.

5.3.2. Treatment Using Submerged Membrane Bio Reactor

Membrane bioreactor is the selective process which is integrated with biological process. It is a combination of both membrane treatment and biological treatment. It is a widely used technique in both industrial and residential treatment system. Membrane bioreactor is an efficient method to remove about 80% anionic surfactants, and it will remove about 87%. COD content and total suspended solids, fecal and total coliform decreased significantly. Nitrogen removal may vary slightly only. Overall, the SMBR satisfy international average values for indoor reuse applications.

SMBR consist of a bioreactor with working volume of 1.0m^3 which also includes grey water flushes and flat plate membrane with pore size of 0.04 mm. Aeration was provided at the base of membrane module. Grey water enters into the system by gravity. Water level is controlled by float switch and suction pump. The treatment unit is also equipped with UV lamp. The UV dose at the maximum flow rate was 40mWs/cm^2. Flow meter is also attached into the system for measuring quantity of grey water. By the experiment studies, results show that on-site MBR systems with daily grey water reclamation of more than 500 liters could be cost-effective (Jabornig (2014)).

5.3.3. Sequencing Batch Reactor (SBR)

Sequencing batch reactor process is based the principle of activated sludge process. A sequencing batch reactor (SBR) was used for a high strength greywater. In the process the sludge rention time is 15 days and hydraulic retention time is 11.7 hours. In the effluent the COD, TP, TN and ammonia removal percentage is 89%, 68%, 35%, and 66% respectively (Bhadiyadra and Vaghani (2015)).

5.3.4. Upflow Anaerobic Sludge Blanket

UASB working principle is based upon the anaerobic process which produce more settleable sludge of active biomass. It gives better performance when we use it with integrated with other treatment systems. In this system the removal efficiency of TSS is 19.3%, COD is 57.8%, BOD is 67.5%, and 83% of oil and grease is removed (Oteng-Peprah et al. (2018)).

5.3.5. Study on Stepped Ecofilter (Vermifilter)

This study was aimed to investigate the hydraulic loading rate (HLR) for the optimal removal efficiency and to analyze the processing performance throughout an entire year. The pilot plant consists of collecting tank and settling and two ecological tanks. A set of improved stepped eco-filter (EF) process based on the CW and the VF was designed above the ground to treat the GW from a farm household in mountainous rural regions. This study observes the GW treatment process in summer and winter based upon the temperature and hydraulic loading rate. The effect of the air temperature on the performance of the EF was examined based on the comparison of the data from the summer and winter in a year, with the temperature 21–30^{0}C in summer and 5–12^{0}C in winter. In the summer, the system removal rate of COD, TN, TP, LAS was 77%, 65%, 45%, 76% and the corresponding value in the winter was 60%, 48%, 41%, 68%, respectively. Therefore, the optimal HLR of the process ought to be in the range of 0.2–0.4 $m^3/(m^2$ day).

5.3.6. Rotating Biological Contactor (RBC) along with Phytotreatment

RBC is a fixed bed reactor type which consists of horizontal rotating shaft with mounted rotating disk. Reactor is partially submerged and microbes are exposed to atmosphere to get oxygen supply for degradation of pollutants and nutrients. Rotating Biological Contactor (RBC) along with a mathematical model was done to analyze the different concentrations of grey water. The RBC system proved to be efficient in BOD removal and it was between 93% - 96% and TSS removal was between 84% - 95% of all concentrations of Influent grey water (Srivastava A. (2018)). High-surface disks mounted on a shaft that continuously rotates in the GW. These disks are only partially immersed in the GW. Disks are covered with microorganisms that degrade organic matter (Maimon and Gross (2018)).

6. Phytotoxicity Test

Phytotoxicity test is carried out to analyze seeds performance and to know whether the grey water is suitable for irrigatn. The phytotoxicity test with seeds of V. radiate and V. mango showed that treated GW sample can be used for irrigation purposes. Particularly, these seeds were used because they are considered as highly prized pulses in the Indian subcontinent. They are highly consumed by the masses hence all the more the reason to carry out the phytotoxicity test on these seeds.

The plants were watered with distilled wter, untreated and treated grey water respectively and it was observed for 30 days. Germination rate was observed in both the experimental plants V. radiate and V. mango in comparison to untreated GW where only 70% growth rate was observed. The root and shoot length was also found to be increased in the presence of treated water sample and thus, it can be concluded that the treated water can be easily used for the cultivation process in India (Singh et al. 2016).

Conclusion

This paper reveals characterization of grey water and discusses about the several technologies for reclamation of grey water through examining the various literatures. Water demand is a serious issue which can be solved by reuse and reclamation technologies. The reclamation of grey water process is a beneficiary solution which can be practised widely for the water demand issues. Reclamation technologies vary depending upon the source and characteristics of grey water. With the declining ground water levels and insufficient surface water available for domestic use, it is the need of the hour to practice grey water treatment and reclamation using the feasible technologies.

REFERENCES

Al-Hamaiedeh, D. H., Bino, M. (2010). Effect of treated grey water reuse in irrigation on soil and plants. *Desalination,* 256: 115–119.

Al-Jayyousi, R. O. (2003). Greywater reuse: towards sustainable water management. *Desalination,* I56:18 l-l 92.

Al-Zouby, Y. J., Al-Zboon, K. K. and Al-Tabbal, J. (2017). Low-cost treatment of grey water and reuse for irrigation of home garden plants. *Environmental Engineering and Management Journal,* 16(2): 351-359.

Andres, G. E., Agullo-Barcelo, M., Bond P., Keller J., Gernjak W. and Radjenovic J. (2018). Hybrid electrochemical-granular activated carbon system for the treatment of greywater. *Chemical Engineering Journal,* 352: 405-412.

Arugam, K., Ghadimi, A., and Chang, L. H., (2018). Design and Optimisation of Home Scale Greywater Recycling Package. *MATEC Web of Conferences,* 152, 02005.

Bhadiyadra, G. J. and Vaghani, V. M. (2015) A review on applicability of photocatalyst titanium dioxide for treatment of greywater. *Int. Journal of Engineering Research and Applications,* 5(3):102-105.

Bhattacharya, P., Sarkar, S., Ghosh, S., Majumdar, S., Mukhopadhyay, A. and Bandyopadhyay, S. (2013). Potential of ceramic microfiltration and ultrafiltration membranes for the treatment of gray water for an effective reuse. *Desalination and Water Treatment,* 51: 4323–4332.

Barısci, S., Sarkka,H., Sillanpaa M. and Dimoglo A. (2015). The treatment of greywater from a restaurant by electrosynthesized ferrate(VI) ion. *Desalination and Water Treatment*, 57(24):11375-11385.

Chen, J., Liao, Z., Lu, S., Hu, G., Liu, Y., Tang, C. (2017). Study on a stepped eco-filter for treating greywater from single farm household. *Applied Water Science,* 7(7): 3849–3857.

Deshmukh, A. V., Hangargekar, P. A. and Paul, D. C. (2017). Treatment and reuse of waste water in coea campus. *International Research Journal of Engineering and Technology,* 04 (8): 241-247.

Ding, G. K. C. (2017). Wastewater Treatment and Reuse-The Future Source of Water Supply. *Encyclopedia of Sustainable Technologies,* Elsevier. 43 - 52.

Dwumfour-Asare, B., Adantey, P., Adantey, B. K. and Appiah-Effah, E. (2017). Greywater characterization and handling practices among urban households in Ghana: the case of three communities in KumasiMetropolis. *Water Science & Technology,* 7(4): 813-822.

Fouad, A. H., Elhefny, M. R. and Hanna, F. H. (2017). Reuse of greywater. *Journal of Applied Sciences Research,* 13(1): 1-9.

Jabornig, S. (2014). Overview and feasibility of advanced grey water treatment systems for single households. *Urban Water Journal*, 11(5): 361–369.

James, D. T. K. and Ifelebuegu, A. O. (2018). Low Cost Sustainable Materials for Grey Water Reclamation. *Water science and technonlgy,* 225: 667-678.

Kuang-Wei, S., Cheng-Wen, W., Jiang, C. S., (2018). Quantitative microbial risk assessment of Greywater on-site reuse. *Civil and Environmental Engineering,* 635: 1507–1519.

Laaffat, J., Aziz, F., Ouazzani, N. and Mandi, L. (2019). Biotechnological approach of greywater treatment and reuse for Landscape irrigation in small communities. *Saudi Journal of Biological Sciences,* 26: 83-90.

Li, F., Wichmann, K., and Otterpohl, R. (2009). Review of the technological approaches for grey water treatment and reuses. *Science of the Total Environment,* 407: 3439–3449.

Maimon, A. and Gross, A. (2018). Greywater: Limitations and perspective. *Current Opinion in Environmental Science & Health,* 2:1–6.

Manouchehri, M. and Kargari, A. (2017). Water recovery from laundry wastewater by the cross flow microfiltration process: A strategy for water recycling in residential buildings. *Journal of Cleaner Production.* 168:227-238.

Moges, E. M., Todt, D., Eregno, E. F. and Heistad, A. (2017). Performance study of biofilter system for on-site greywater treatmentat cottages and small households. *Ecological Engineering,* 105:118–124.

Oteng-Peprah, M., Acheampong, A. M. and DeVries, K. N. (2018). Greywater, Characteristics, Treatment Systems, Reuse Strategies and User Perception- a review. *Water Air Soil Pollut,* 229(255):1-16.

Pidou, M., Avery, L., Stephenson, O., Jeffrey, P., Parsons, A. S., Liu, S., Memon, A. F., Jefferson, B. (2008). Chemical solutions for greywater recycling. *Chemosphere,* 71: 147–155.

Ramprasada, C., Smith, C. S., Memon, F. A. and Philip, L. (2017) Removal of chemical and microbial contaminants from greywater using a novel constructed wetland. *Ecological Engineering,* 106: 55–65.

Reichman, M. S. and Wightwick, M.A. (2013). Impacts of standard and 'low environmental impact' greywater, irrigation on soil and plant nutrients and ecology. *Applied Soil Ecology,* 72: 195– 202.

Singh, S., Pradhan N., Ojha N., Roy B. and Bose S. (2016). Grey Water treatment and its application in cultivation of plants. *Asian Jr. of Microbiol. Biotech. Env. Sc.*, 18(4): 1043-1053.

Song, Y., Men, B., Wang, D., & Ma, J. (2017). On-line batch production of ferrate with an chemical method and its potential application for greywater recycling with Al(III) salt. *Journal of Environmental Sciences*, 52, 1–7.

Srivastava, A. (2018). Performance evaluation of grey water treatment using rotating biological contactor (rbc) along with phytotreatment and future scope for use as drinking water. *International Journal of Advance Research,* 4(2): 1289-1297.

Winwarda, P. G., Averyb, M. L., Stephensona, T., Jeffersona, B. (2008). Chlorine disinfection of grey water for reuse: Effect of organics and particles. *Water research,* 42: 483 – 491.

Zhu, J., Wagner, M., Cornel, P., Chen, H. and Dai, X. (2018). Feasibility of on-site grey-water reuse for toilet flushing in China. *Journal of water reuse and desalination,* 8(1),1-13.

INDEX

A

acetate, 85, 86, 155
activated sludge, vii, ix, x, 2, 7, 31, 34, 37, 40, 41, 46, 47, 52, 53, 54, 55, 57, 58, 61, 62, 67, 68, 69, 71, 73, 86, 89, 93, 102, 117, 118, 119, 134, 147, 158, 168, 169, 213
adsorption, 25, 49, 50, 52, 56, 65, 74, 75, 146, 148, 212
advanced oxidation, 146, 209, 210
advanced oxidation processes, 146, 177
aeration, ix, 2, 36, 38, 41, 44, 45, 71, 74, 114, 138, 149, 156, 157, 161, 213
aerobic granules, vi, xi, 145, 146, 148, 155, 156, 157, 159, 160, 161, 165, 168, 169, 173, 179
agriculture, vii, viii, 2, 3, 4, 5, 8, 10, 12, 14, 15, 18, 21, 22, 23, 25, 26, 27, 28, 29, 30, 39, 92, 96, 131, 133, 135, 168, 202
algal cell immobilization technique, 152, 167
amides, 150
ammonia, xi, 38, 50, 90, 106, 113, 114, 145, 146, 152, 163, 176, 177, 214
AnSBBR, xii, 182, 184, 185, 186, 189, 193, 197, 198, 199, 200
aromatic, 146, 148, 151
asphaltenes, 150

B

bacilli, 155
baffles, 156, 161
benzene, 150, 151, 166
benzo (a) pyrene, 151
bioethanol, v, vii, xi, xii, 95, 96, 98, 99, 100, 101, 106, 110, 111, 112, 121, 122, 123, 127, 128, 130, 131, 133, 135, 136, 137, 138, 139, 140, 141, 142, 143, 182
biofuels, xi, 95, 96, 98, 100, 106, 109, 123, 133, 139, 140, 144
biogas, v, vii, xi, 95, 96, 97, 101, 102, 103, 104, 105, 106, 112, 113, 114, 117, 118, 119, 120, 121, 123, 126, 127, 128, 130, 131, 132, 134, 135, 136, 137, 138, 139, 140, 141, 142, 143, 144, 181, 182, 183, 184, 185, 186, 187, 188, 190, 192, 195, 196
biological oxygen demand (BOD), 54, 73, 108, 123, 124, 125, 151, 152, 204, 206, 209, 210, 214, 215
biological procedures, 32
bioremediation, 62, 148, 152, 162

bitumen, 150
BTEX, xi, 145, 151
bulk volume, 153

C

carboxylic acids, 150
centrifuge, 146
chemical oxygen demand (COD), 17, 37, 54, 55, 56, 57, 58, 88, 112, 115, 116, 117, 118, 119, 120, 121, 123, 124, 132, 149, 151, 152, 157, 158, 165, 179, 187, 188, 190, 191, 194, 195, 204, 205, 209, 210, 212, 213, 214
circular economy, vii, xi, 96, 103, 104, 106, 113, 120, 127, 137
clean energy, 75, 96
coagulation, 38, 72, 122, 146, 163, 171, 172, 177, 205, 209, 210
cocci, 155, 161
co-digestion, vi, xii, 102, 131, 135, 181, 182, 183, 185, 187, 190, 192, 193, 194, 195, 196, 197, 198, 199, 200
constructed wetland, 11, 27, 28, 162, 167, 205, 212, 218
crude oil, 146, 150
cyanides, xi, 145, 152

D

density, vii, xi, 46, 72, 77, 79, 146, 153, 156, 158, 160, 161, 186, 210
depollution, xi, 96
drag force, 158

E

eco-friendly, xi, 96
energy production, x, xi, 70, 78, 91, 93, 96, 130, 183, 198

F

fatty acids, 150, 190
filamentous, 46, 155, 161
flocculation, 38, 73, 163, 210
flocs, 38, 41, 147, 154

fl

flotation, 146

F

flotation, 163, 171, 172
fluffy, 154
freezing, 146

G

glucose, 86, 98, 112, 155, 161
glycerin, vi, xii, 181, 182, 183, 185, 187, 188, 189, 190, 191, 192, 193, 194, 195, 196, 198, 199
granular activated carbon (GAC), 119, 211, 216
granular sequencing batch reactor (GSBR), vi, 146, 148, 149, 151, 152, 153, 155, 156, 157, 158, 159, 160, 161
granules, vii, xi, 145, 148, 149, 151, 153, 154, 155, 156, 157, 158, 159, 160, 164, 168, 170, 174, 204
gravity, ix, 2, 73, 86, 158, 213
green economy, xi, 96
grey water, vi, viii, xiii, 201, 202, 203, 204, 205, 206, 207, 209, 210, 211, 212, 213, 215, 216, 217, 218, 219

H

heavy metals, xi, 11, 14, 33, 61, 75, 114, 120, 145, 146, 148, 151, 152, 167, 172, 204
hydrocarbon, 150, 175
hydrodynamic shear force, 148, 156, 161

I

inhibitory concentration, 152
inorganic contaminants, 151, 152
integrity coefficient, 159, 160

L

lead, 36, 151, 156, 161, 173, 207, 211

M

membrane bioreactors, 205
mercaptans, 146, 150
mercury, 151
methane, viii, xi, 43, 46, 97, 98, 101, 102, 120, 122, 124, 126, 130, 136, 139, 143, 181, 182, 183, 185, 187, 190, 192, 193, 195, 196, 197, 198, 199
methyl tetra-butyl ether, 151
microfiltration, 123, 129, 202, 209, 216, 218
morphology, 154, 159
moving bed biofilm reactor, 147, 171
municipal wastewater, v, 3, 6, 11, 27, 28, 29, 30, 31, 42, 54, 60, 61, 62, 68, 84, 88, 89, 90, 132, 144, 203

N

naphthenic acids, 150
naphthenics, 150
nickel, 151, 176
nitrification, 38, 152, 167
nitrogen, viii, 1, 2, 3, 6, 26, 38, 46, 54, 55, 58, 72, 80, 81, 93, 113, 114, 115, 116, 123, 136, 149, 150, 165, 168, 183, 203, 205, 212, 213
nutrient, 8, 9, 10, 26, 27, 29, 47, 93, 101, 104, 112, 113, 114, 115, 118, 119, 120, 121, 132, 138, 140, 142, 148, 160, 169, 183, 192, 205, 206

O

oily wastewater, 146, 147, 161, 163
organic contaminants, 151, 167
organic loading, xi, 146, 148, 174, 187, 188, 193, 198, 200

P

paraffinics, 150
performance, v, vii, viii, x, xi, 29, 31, 32, 34, 43, 52, 54, 55, 56, 57, 79, 80, 81, 82, 87, 88, 91, 102, 118, 119, 128, 129, 132, 139, 142, 146, 165, 167, 169, 170, 174, 175, 177, 179, 187, 190, 192, 193, 194, 198, 199, 207, 210, 214, 215, 218
petroleum, vi, viii, xi, 134, 142, 145, 146, 150, 151, 152, 155, 156, 157, 159, 160, 161, 162, 163, 164, 165, 166, 167, 173, 174, 175, 179
petroleum wastewater, vi, xi, 145, 150, 155, 156, 157, 159, 160, 161, 162, 164, 166, 173, 179
phenol, xi, 122, 145, 146, 150, 151, 152, 167, 171, 174
phenolic compounds, 150
phosphorus, viii, 1, 2, 3, 6, 10, 38, 46, 54, 72, 113, 114, 123, 150, 203, 204
physical strength, viii, xi, 146, 153, 159
physico-chemical treatment, 202
phytotoxicity, 11, 202, 215

pollution, vii, viii, x, 1, 2, 3, 22, 25, 32, 33, 34, 36, 41, 44, 54, 55, 56, 58, 60, 62, 67, 91, 146, 148, 160, 162, 164, 166, 168, 172, 175, 182, 183, 203, 205, 206
purification, 29, 33, 35, 36, 38, 40, 41, 42, 45, 46, 47, 54, 57, 58, 59, 64, 72, 99, 143, 153
pyridines, 150

R

recycling, vii, viii, ix, xi, xii, 8, 25, 31, 32, 90, 96, 97, 103, 104, 108, 110, 112, 121, 122, 123, 126, 127, 129, 130, 131, 133, 137, 139, 143, 201, 205, 208, 211, 216, 218
renewable fuels, xi, 95
reuse, iv, vi, vii, viii, ix, xi, xiii, 4, 9, 13, 17, 18, 23, 25, 26, 29, 30, 31, 33, 34, 39, 59, 61, 64, 68, 74, 93, 96, 100, 101, 104, 105, 107, 108, 109, 110, 112, 121, 122, 125, 126, 127, 128, 129, 130, 131, 137, 140, 164, 172, 175, 201, 202, 203, 213, 216,217, 218, 219
Reynolds number, 158

S

Saudi Arabia, vii, ix, 1, 2, 4, 15, 16, 17, 22, 25, 26, 27, 28, 29
scanning electron microscopy (SEM), 154
seed sludge, 148, 154, 155, 157
sequencing batch reactor (SBR), xi, 133, 139, 145, 147, 148, 149, 152, 153, 161, 165, 166, 167, 168, 170, 198, 200, 205, 213
sludge, vii, xi, 2, 7, 8, 9, 15, 21, 24, 26, 27, 35, 36, 37, 38, 41, 42, 44, 45, 46, 49, 55, 59, 66, 73, 88, 97, 101, 103, 112, 119, 131, 132, 139, 144, 146, 147, 149, 152, 153, 154, 156, 159, 160, 163, 164, 165, 168, 169, 170, 175, 187, 200, 214
sludge bulking, 147
sludge volume index, vii, xi, 146, 149, 156, 169
specific gravity, 156
specific oxygen uptake rate, 152
Stokes equation, 158

T

thiols, 150
thiophenes, 150
total dissolved solids (TDS), 6, 15, 16, 17, 122, 152
transmission electron microscope (TEM), 154, 155

U

ultraviolet irradiation, 39
up-flow anaerobic sludge blanket (UASB), 103, 128, 147, 168, 187, 200, 214

V

vinasse, vi, xii, 97, 108, 110, 122, 124, 126, 181, 182, 183, 184, 185, 187, 188, 189, 190, 192, 193, 194, 195, 196, 197, 198, 199, 200

W

wastewater quality, 15, 65
wastewater recycling, 32, 104
water uses, 101
whey, vi, xii, 116, 181, 182, 183, 185, 187, 188, 189, 190, 192, 193, 194, 195, 196, 197, 198, 199, 200